GLENCOE MATHEMATICS

Noteables™

Interactive Study Notebook with Foldables™

Algebra

Concepts and Applications

Contributing Author
Dinah Zike

Foldables™

Consultant
Douglas Fisher, PhD
Director of Professional Development
San Diego State University
San Diego, CA

McGraw Hill Glencoe

New York, New York Columbus, Ohio Chicago, Illinois Peoria, Illinois Woodland Hills, California

Glencoe

The McGraw·Hill Companies

Send all inquiries to:
The McGraw-Hill Companies
8787 Orion Place
Columbus, OH 43240-4027

ISBN: 0-07-872985-8

Algebra: Concepts and Applications (Student Edition)
Noteables™: Interactive Study Notebook with Foldables™

3 4 5 6 7 8 9 10 024 09 08 07 06

Contents

CHAPTER 8

CHAPTER 9

CHAPTER 10

CHAPTER 11

CHAPTER 12

CHAPTER 13

CHAPTER 14

CHAPTER 15

Organizing Your Foldables

Make this Foldable to help you organize and store your chapter Foldables. Begin with one sheet of 11" × 17" paper.

STEP 1 **Fold**
Fold the paper in half lengthwise. Then unfold.

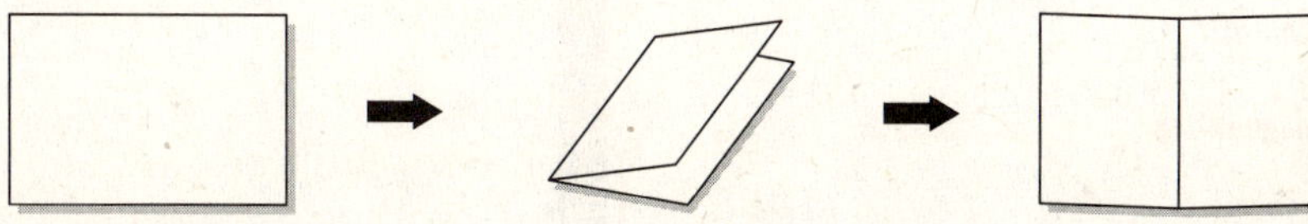

STEP 2 **Fold and Glue**
Fold the paper in half widthwise and glue all of the edges.

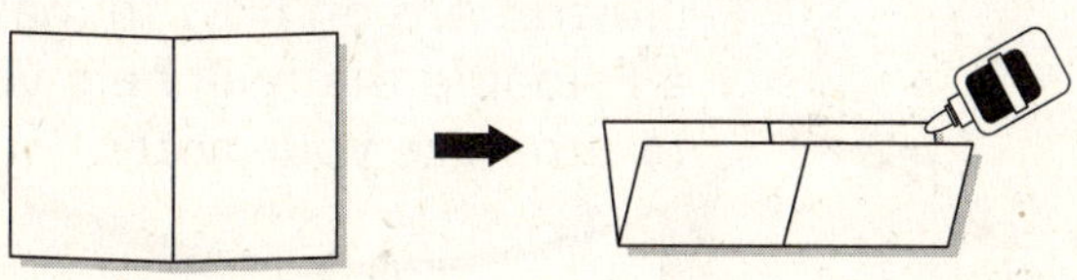

STEP 3 **Glue and Label**
Glue the left, right, and bottom edges of the Foldable to the inside back cover of your Noteables notebook.

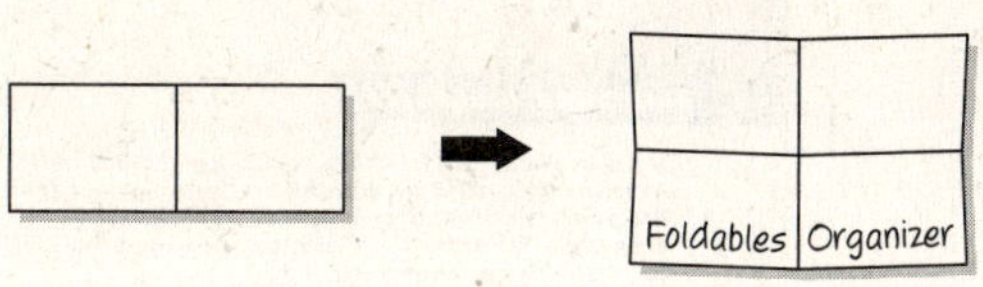

Reading and Taking Notes As you read and study each chapter, record notes in your chapter Foldable. Then store your chapter Foldables inside this Foldable organizer.

Using Your Noteables™ Interactive Study Notebook with Foldables™

This note-taking guide is designed to help you succeed in *Algebra: Concepts and Applications*. Each chapter includes:

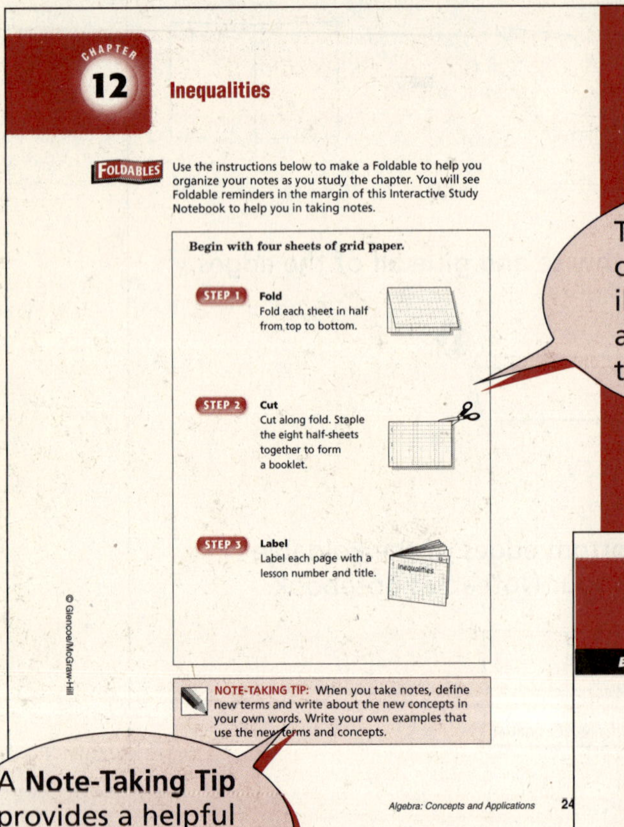

Chapter 12

Inequalities

FOLDABLES Use the instructions below to make a Foldable to help you organize your notes as you study the chapter. You will see Foldable reminders in the margin of this Interactive Study Notebook to help you in taking notes.

Begin with four sheets of grid paper.

STEP 1 **Fold** Fold each sheet in half from top to bottom.

STEP 2 **Cut** Cut along fold. Staple the eight half-sheets together to form a booklet.

STEP 3 **Label** Label each page with a lesson number and title.

NOTE-TAKING TIP: When you take notes, define new terms and write about the new concepts in your own words. Write your own examples that use the new terms and concepts.

© Glencoe/McGraw-Hill

Algebra: Concepts and Applications 24

The **Chapter Opener** contains instructions and illustrations on how to make a Foldable that will help you to organize your notes.

A **Note-Taking Tip** provides a helpful hint you can use when taking notes.

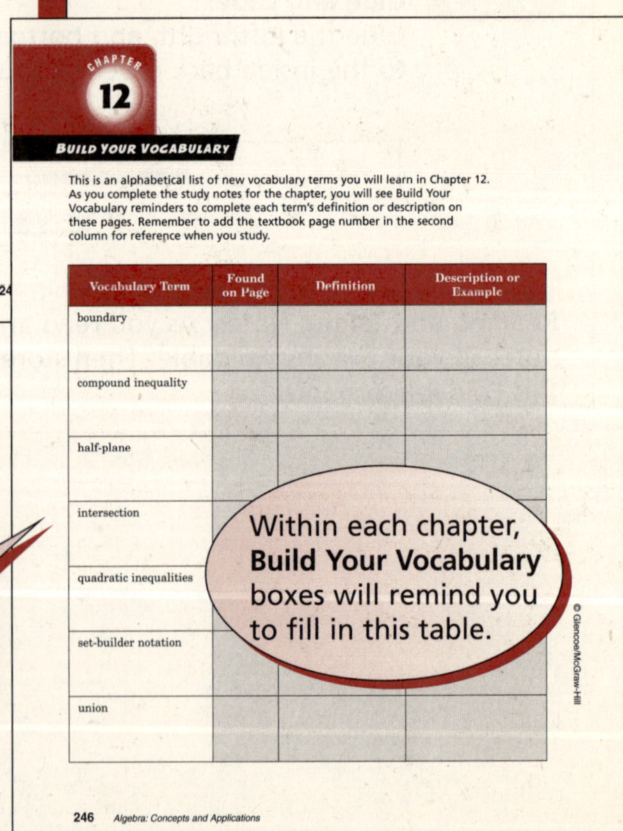

Chapter 12

BUILD YOUR VOCABULARY

This is an alphabetical list of new vocabulary terms you will learn in Chapter 12. As you complete the study notes for the chapter, you will see Build Your Vocabulary reminders to complete each term's definition or description on these pages. Remember to add the textbook page number in the second column for reference when you study.

Vocabulary Term	Found on Page	Definition	Description or Example
boundary			
compound inequality			
half-plane			
intersection			
quadratic inequalities			
set-builder notation			
union			

246 *Algebra: Concepts and Applications*

© Glencoe/McGraw-Hill

The **Build Your Vocabulary** table allows you to write definitions and examples of important vocabulary terms together in one convenient place.

Within each chapter, **Build Your Vocabulary** boxes will remind you to fill in this table.

12-5 Solving Compound Inequalities

What You'll Learn

- Solve compound inequalities.

Build Your Vocabulary (page 246)

Two or more inequalities that are connected by the words ☐ or ☐ form a **compound inequality**.

An **intersection** is the set of elements common to ☐ inequalities.

A **union** is the set of elements in each of ☐ inequalities.

Lessons cover the content of the lessons in your textbook. As your teacher discusses each example, follow along and complete the **fill-in boxes.** Take notes as appropriate.

Foldables

Organize It

Summarize the difference between "intersection" and "union" under the tab for Lesson 12-5. Then, give examples of when to use each one.

EXAMPLE

1. **Write $x < 0$ and $x \geq -3$ as a compound inequality without using *and*.**

$x < 0$ and $x \geq -3$ can be written as ☐ $> x \geq$ ☐ or ☐ $\leq x <$ ☐.

Your Turn Write $x > -2$ and $x \leq 5$ as a compound inequality without using *and*.

EXAMPLE

2. **A veterinarian has a scale for weighing dogs and ca[ts] that weigh more than 10 pounds *but no more than 65 pounds*. The weights w that can be measured on ... can be written as $10 < w \leq 65$. Graph the solu[tion] ...**

...quality using ☐. $10 < w$ ...

Foldables feature reminds you to take notes in your Foldable.

© Glencoe/McGraw-Hill

...gebra: Concepts and Applications

Examples parallel the examples in your textbook.

12-4

EXAMPLE

4. **Karl's point totals in the first four of five basketball games were 15, 12, 19, and 18. How many points t must he score in the fifth game to have a mean point total of more than 16?**

The sum of Karl's points, divided by ☐, will give the mean score. The mean must be more than ☐.

$$\frac{15 + 12 + 19 + 18 + t}{\square} > \square$$

$$\square\left(\frac{15 + 12 + 19 + 18 + t}{\square}\right) > \square(16)$$ Multiply each side by ☐.

$$15 + 12 + 19 + 18 + t > \square$$

$$\square + t > 80$$

$$64 + t - \square > 80 - \square$$ Subtract.

$$t > \square$$

Karl must score more than ☐ points in the fifth game to have a mean point total of more than ☐.

Your Turn Lien's score on the first four of five 100-point tests were 82, 85, 95, and 91. What score s on the fifth test will give her a mean score of at least 90 for all five tests?

Homework Assignment

Page(s):

Exercises:

© Glencoe/McGraw-Hill

256 Algebra: Concepts and Applications

Your Turn Exercises allow you to solve similar exercises on your own.

Chapter 12 Bringing It All Together

Study Guide

Foldables	Vocabulary Puzzlemaker	Build Yo[ur] Vocabu[lary]
Use your **Chapter 12 Foldable** to help you study for your chapter test.	To make a crossword puzzle, word search, or jumble puzzle of the vocabulary words in Chapter 12, go to: www.glencoe.com/sec/math/t_resources/free/index.php	You can use yo[ur] **Vocabulary Buil[der]** (page 246) to h[elp] the puzzle.

12-1 Inequalities and Their Graphs

Write the letter of the graph that matches each inequality.

1. $x \leq -1$ ☐
2. $x \geq -1$ ☐
3. $x < -1$ ☐
4. $x > -1$ ☐

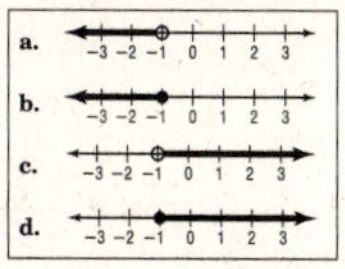

12-2 Solving Addition and Subtraction Inequalities

Write an inequality for each statement. Then solve.

5. A number subtracted from 21 is no less than −2.
6. A number added to 12 is a minimum of −1.
7. 5 more than a number is at least 15.
8. 18 less than a number is at most 45.

© Glencoe/McGraw-Hill

Algebra: Concepts and Applications 265

Bringing It All Together Study Guide reviews the main ideas and key concepts from each lesson.

NOTE-TAKING TIPS

Your notes are a reminder of what you learned in class. Taking good notes can help you succeed in mathematics. The following tips will help you take better classroom notes.

- Before class, ask what your teacher will be discussing in class. Review mentally what you already know about the concept.
- Be an active listener. Focus on what your teacher is saying. Listen for important concepts. Pay attention to words, examples, and/or diagrams your teacher emphasizes.
- Write your notes as clear and concise as possible. The following symbols and abbreviations may be helpful in your note-taking.

Word or Phrase	Symbol or Abbreviation	Word or Phrase	Symbol or Abbreviation
for example	e.g.	not equal	$\neq$
such as	i.e.	approximately	$\approx$
with	w/	therefore	$\therefore$
without	w/o	versus	vs
and	+	angle	$\angle$

- Use a symbol such as a star (★) or an asterisk (*) to emphasis important concepts. Place a question mark (?) next to anything that you do not understand.
- Ask questions and participate in class discussion.
- Draw and label pictures or diagrams to help clarify a concept.
- When working out an example, write what you are doing to solve the problem next to each step. Be sure to use your own words.
- Review your notes as soon as possible after class. During this time, organize and summarize new concepts and clarify misunderstandings.

Note-Taking Don'ts

- **Don't** write every word. Concentrate on the main ideas and concepts.
- **Don't** use someone else's notes as they may not make sense.
- **Don't** doodle. It distracts you from listening actively.
- **Don't** lose focus or you will become lost in your note-taking.

CHAPTER 1

The Language of Algebra

FOLDABLES™ Use the instructions below to make a Foldable to help you organize your notes as you study the chapter. You will see Foldable reminders in the margin of this Interactive Study Notebook to help you in taking notes.

Begin with four sheets of $8\frac{1}{2}$" × 11" paper.

STEP 1 **Stack** sheets of paper with edges $\frac{3}{4}$ inch apart.

STEP 2 **Fold** up bottom edges. All tabs should be the same size.

STEP 3 **Staple** along the fold.

STEP 4 **Label** the tabs with topics from the chapter.

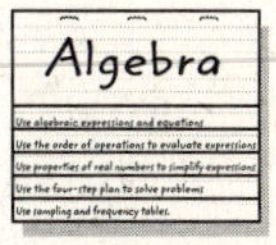

NOTE-TAKING TIP: When you take notes, it is important to record important concepts. Be sure to refer to your notes when reviewing for tests.

CHAPTER 1

BUILD YOUR VOCABULARY

This is an alphabetical list of new vocabulary terms you will learn in Chapter 1. As you complete the study notes for the chapter, you will see Build Your Vocabulary reminders to complete each term's definition or description on these pages. Remember to add the textbook page number in the second column for reference when you study.

Vocabulary Term	Found on Page	Definition	Description or Example
algebraic expression [al-juh-BRAY-ik]			
coefficient [CO-i-FISH-unt]			
conclusion			
conditional			
counterexample			
data			
deductive reasoning [dee-DUK-tiv]			
equation [EE-KWAY-zhun]			
equivalent expressions [ee-KWIV-a-lunt]			
evaluate [ee-val-yoo-WAYT]			
factors			
formula [FOR-myu-la]			
frequency table			
histogram			

Vocabulary Term	Found on Page	Definition	Description or Example
hypothesis [hi-PA-the-sis]			
if-then statement			
inductive reasoning [in-DUK-tiv]			
like terms			
line graph			
numerical expression [noo-MARE-ik-ul]			
order of operations			
population			
product			
quotient			
sample			
sampling			
simplest form			
simplify			
stem-and-leaf plot			
term			
variable [VARE-ee-a-bul]			
whole numbers			

1–1 Writing Expressions and Equations

What You'll Learn

- Translate words into algebraic expressions and equations.

Build Your Vocabulary (pages 2–3)

An **algebraic expression** contains ______ and ______ along with +, −, ×, and/or ÷.

A **variable** is a ______ used to represent an unknown number.

A **numerical expression** contains ______ along with +, −, ×, and/or ÷.

A **factor** is a quantity that is being ______.

A **product** is the result of ______ that are multiplied.

An **equation** is a mathematical ______ that contains an equals sign (=).

Foldables™

Organize It

In your notes, write several words or phrases for each of the operations, addition, subtraction, multiplication, and division.

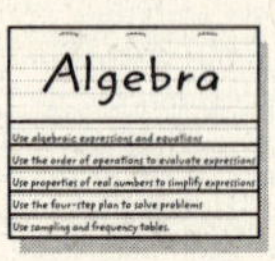

Examples

Write an algebraic expression for each verbal expression.

1. **the sum of p and 12**

2. **the product of k and q**

Your Turn **Write an algebraic expression for each verbal expression.**

a. b decreased by 6

b. the product of 9 and z

EXAMPLES

A python eats 4 pounds of meat each month.

3 **Write a numerical expression to represent the amount it eats in 5 months.**

The numerical expression is ☐ · ☐ or 4(5).

4 **Write an algebraic expression to represent the amount it eats in d months.**

The algebraic expression is $4 \cdot d$ or ☐.

Your Turn **A spider has eight legs. Write an expression for each situation.**

a. the number of legs on 3 spiders

☐

b. the number of legs on m spiders

☐

EXAMPLES

Write a verbal expression for each algebraic expression.

5 **37 + s**

the ☐ of ☐ and ☐

37 ☐ by s

6 **5(b − 3)**

5 ☐ the difference of ☐ and ☐

5 ☐ by the ☐ of b and 3

Your Turn **Write a verbal expression for each algebraic expression.**

a. $x \div 7$

☐

b. $4 + (2n)$

☐

WRITE IT

Write one other verbal expression for Examples 5 and 6.

______

______

______

______

______

______

EXAMPLES

Write an equation for each sentence.

7 **The quotient of t and 8 equals 20.**

t [ ] 8 [ ] 20 or [ ] = 20

8 **Seven less than three times g is 31.**

[ ] – [ ] = 31

Your Turn **Write an equation for each sentence.**

a. A number x increased by 12 is the same as 27

[ ]

b. The product of 8 and d decreased by 9 equals 23.

[ ]

EXAMPLES

Write a sentence for each equation.

9 $j + 4 = 21$

[ ] more than j is [ ] 21.

10 $3z - 12 = 11$

The product of [ ] decreased by 12 [ ] 11.

Your Turn **Write a sentence for each equation.**

a. $r - 10 = 5$

[ ]

b. $\frac{n}{3} + 6 = 30$

[ ]

HOMEWORK ASSIGNMENT

Page(s):

Exercises:

1–2 Order of Operations

WHAT YOU'LL LEARN

- Use the order of operations to evaluate expressions.

KEY CONCEPT

Order of Operations

1. Find the values of expressions inside grouping symbols, such as parentheses ( ), brackets [ ], and as indicated by fraction bars.
2. Do all multiplications and/or divisions from left to right.
3. Do all additions and/or subtractions from left to right.

FOLDABLES In your notes, summarize the Order of Operations.

EXAMPLES

Find the value of each expression.

1 $14 + 10 \div 2$

$14 + 10 \div 2 = 14 +$ ____ ____ 10 and 2.

$=$ ____ ____ 14 and 5.

2 $4 \times (6 + 7)$

$4 \times (6 + 7) = 4 \times 13$ ____ 6 and 7.

$= 52$ ____ 4 and 13.

Your Turn **Find the value of each expression.**

a. $(2 + 4) \div 2$

____

b. $3 \cdot 5 - 2 \cdot 1$

____

EXAMPLES

Name the property of equality shown by each statement.

3 **If $k = 7$, then $k + 3 = 7 + 3$.**

____ Property of Equality

4 **If $a + 4 = 9$, then $9 = a + 4$.**

____ Property of Equality

Your Turn **Name the property of equality shown by each statement.**

a. If $a + 2 = 5$ and $5 = 3 + 2$, then $a + 2 = 3 + 2$.

____

b. If $x = 7$, then $10 - x = 10 - 7$.

____

KEY CONCEPTS

Properties of Equality
Substitution If $a = b$, then a may be replaced by b.

Reflexive $a = a$

Symmetric If $a = b$, then $b = a$.

Transitive If $a = b$, and $b = c$, then $a = c$.

Properties of Numbers
Additive Identity When 0 is added to any number a, the sum is a.

Multiplicative Identity When a number a is multiplied by 1, the product is a.

Multiplicative Property of Zero If 0 is a factor, the product is 0.

EXAMPLE

5 Find the value of $[25 + 8(12 - 11)] \div 11$. Identify the property used in each step.

$[25 + 8(12 - 11)] \div 11$

$= (25 + 8(\quad)] \div 11$	______ Property
$= (25 + \quad) \div 11$	______ Identity
$= \quad \div 11$	______ Property
$= 3$	______ Property

Your Turn Find the value of $30 \div (6 + 4) \cdot 1 + 5$. Identify the property used in each step.

BUILD YOUR VOCABULARY (page 2)

To **evaluate** an expression is to find the ______ of an expression by replacing the variables with numbers.

EXAMPLE

6 Evaluate $xy + 8$ if $x = 4$ and $y = 3$.

$xy + 8 = \quad + 8$	Replace x with 4 and y with 3.
$= \quad + 8$	Substitution Property
$= 20$	Substitution Property

Your Turn **Evaluate each expression if $a = 7$ and $b = 1$.**

a. $10 - a \div b$

b. $(a + 2) + 4b$

HOMEWORK ASSIGNMENT

Page(s):

Exercises:

1–3 Commutative and Associative Properties

WHAT YOU'LL LEARN

- Use the commutative and associative properties to simplify expressions.

KEY CONCEPTS

Commutative Property of Addition and Multiplication The order in which two numbers are added or multiplied does not change their sum or product.

Associative Property of Addition and Multiplication The way in which three numbers are grouped when they are added or multiplied does not change their sum or product.

EXAMPLES

Name the property shown by each statement.

1 $8 + (3 + 4) = (3 + 4) + 8$

______ Property of ______

2 $7 \cdot (8 \cdot k) = (7 \cdot 8) \cdot k$

______ Property of ______

Your Turn **Name the property shown by each statement.**

a. $5 \cdot 4 \cdot 3 = 5 \cdot 3 \cdot 4$

b. $a + (2 + 8) = (a + 2) + 8$

BUILD YOUR VOCABULARY (page 3)

To **simplify** an expression, ______ all parentheses first and then add, subtract, ______, or divide.

EXAMPLE

3 **Simplify the expression $(4 \cdot m) \cdot 9$. Identify the properties used in each step.**

$(4 \cdot m) \cdot 9$

$= (m \cdot \_\_\_) \cdot 9$ ______ Property (×)

$= m \cdot (4 \cdot \_\_\_)$ ______ Property (×)

$= m \cdot \_\_\_$ ______

$= \_\_\_$ ______ Property (×)

Your Turn Simplify the expression $(12 + z) + 7$. Identify the properties used in each step.

BUILD YOUR VOCABULARY (pages 2–3)

Whole numbers are the numbers 0, 1, 2, 3, 4, and so on.

A **counterexample** is an ______ that shows that the statement is ______.

EXAMPLE

4 State whether the statement *Subtraction of whole numbers is commutative* is *true* or *false*. If false, provide a counterexample.

Write two subtraction expressions using the Commutative Property and check to see whether they are equal.

$7 - 5 = 5 - 7$ Evaluate each expression separately.

$7 - 5 = 5 - 7$ $7 - 5 =$ ______ and $5 - 7 =$ ______

We found a ______, so the statement is ______.

KEY CONCEPT

Closure Property of Whole Numbers Because the sum or product of two whole numbers is also a whole number, the set of whole numbers is closed under addition and multiplication.

Your Turn State whether the statement *Subtraction of whole numbers is associative* is *true* or *false*. If false, provide a counterexample.

HOMEWORK ASSIGNMENT

Page(s):

Exercises:

1–4 Distributive Property

WHAT YOU'LL LEARN

- Use the Distributive Property to evaluate expressions.

KEY CONCEPTS

Distributive Property
For any numbers a, b, and c, $a(b + c) = ab + ac$ and $a(b - c) = ab - ac$.

FOLDABLES

ORGANIZE IT

In your notes, write three examples of terms and three examples of items that are not terms and label each group. Circle the coefficient in each term.

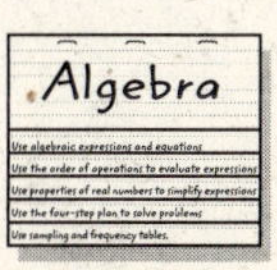

EXAMPLES

Simplify each expression.

1 $\mathbf{5(2 + m)}$

$5(2 + m) = (5 \cdot \underline{\quad}) + (5 \cdot \underline{\quad})$ Distributive Property

$= \underline{\quad} + 5m$ Substitution Property

2 $\mathbf{3(4x + 2)}$

$3(4x + 2) = (3 \cdot \underline{\quad}) + (3 \cdot \underline{\quad})$ Distributive Property

$= \underline{\quad} + 6$ Substitution Property

Your Turn **Simplify each expression.**

a. $7(b + 3)$

b. $3(4t + 8)$

BUILD YOUR VOCABULARY (pages 2–3)

A **term** is a number, ________, product, or quotient of ________ and variables.

The numerical part of a ________ that contains a variable is a **coefficient**.

Like terms are terms that contain the same ________ such as 2*a* and 5*a*.

Expressions whose ________ are the same are **equivalent expressions.**

An algebraic expression is in **simplest form** when it has no like terms and no parentheses.

EXAMPLES

3 **Simplify each expression.**

$8p - 5p$

$8p - 5p = \square p$ — Distributive Property

$= \square$ — Substitution Property

4 **$10k + 6m - 5k + 2m$**

$10k + 6m - 5k + 2m$

$= 10k - \square + 6m + \square$ — Comm. Property (+)

$= (10k - \square) + (6m + \square)$ — Assoc. Property (+)

$= (\square)k + (\square)m$ — Distributive Property

$= \square + \square$ — Substitution Property

Your Turn **Simplify each expression.**

a. $3d + 5d$

b. $6n + 7m - 4n$

HOMEWORK ASSIGNMENT

Page(s):

Exercises:

1–5 A Plan for Problem Solving

WHAT YOU'LL LEARN

- Use a four-step plan to solve problems.

FOLDABLES™

ORGANIZE IT

In your notes, list the seven Problem-Solving Strategies.

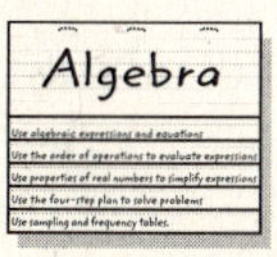

REMEMBER IT

Always check to make sure the answer is reasonable.

EXAMPLE

1 **Suppose you deposit $350 into an account that pays 2% interest. How much money would you have in the account after five years?**

EXPLORE You know the amount of money deposited, the interest rate, and the time.

You need to find the amount of money, including interest, at the end of five years

PLAN Use the formula $I = prt$ and substitute the known values. Add this amount to the original deposit.

Estimate: 1% of $350 is ______. So, 2% of $350 is about 2 × ______ or ______ per year. This will be ______ in five years. You should have about ______ + ______ = ______ in five years.

SOLVE $I = prt$ Interest formula

$I =$ ______ · ______ · ______ $p =$ ______, $r =$ ______, $t =$ ______

$I =$ ______

You will earn ______ in interest, so the total amount after five years is ______ + ______ or ______.

EXAMINE Since ______ is the same as the ______, the answer is reasonable.

Your Turn Suppose you deposit $270 into an account that pays 3% interest. How much money would you have in the account after ten years?

EXAMPLE

2 How many ways can you make 50¢ using quarters, dimes, and nickels?

EXPLORE You need to know how many ways you can make ______ without using ______.

PLAN Make a chart listing every possible combination.

SOLVE

Coin	Number									
Quarters	2	1	1	1	0	0	0	0	0	0
Dimes	0	2	1	0	5	4	3	2	1	0
Nickels										

There are ______ ways to make ______.

EXAMINE Check that each combination totals 50¢ and that there are no other possible combinations.

Your Turn How many ways can you make 30¢ using quarters, dimes, and pennies?

HOMEWORK ASSIGNMENT

Page(s):

Exercises:

1–6 Collecting Data

What You'll Learn

- Collect and organize data using sampling and frequency tables.

Build Your Vocabulary (pages 2–3)

Sampling is a convenient way to ________ **data**, or information, so that ________ can be made about a **population**.

A **sample** is a small ________ used to represent a much larger ________.

Example

1 **One hundred cable-television subscribers are surveyed to find how much time the average American spends reading. Is this a good sample? Explain.**

No; many of those surveyed would prefer ________ to reading.

Key Concept

Sampling Criteria A good sample is representative of the larger population, selected at random, and large enough to provide accurate data.

Your Turn Twenty-five hundred people across the state of Georgia were randomly surveyed to find the average level of completed education. Is this a good sample? Explain.

____

Build Your Vocabulary (page 2)

A **frequency table** is a table that uses tally marks to ________ and display the ________ of events.

EXAMPLE

2 **Make a frequency table to organize the data in the chart. Use intervals of 5.**

Record High Temperatures for Selected U.S. States (°C)				
44	38	53	49	57
48	41	43	41	43
45	38	48	47	47
48	49	46	46	41

STEP 1 Make a table with three ________. Add a title.

STEP 2 Use ________ so there are fewer categories.

In this case, the intervals are of size ____.

STEP 3 Use ________ marks to record the times in each ________.

STEP 4 Count the tally marks in each row and record this number in the ________ column.

Record High Temperatures for Selected U.S. States (°C)		
________	________	________
________	||	2
40–44	________	________
45–49	________	________
________	|	1
55–59	|	________

Your Turn A museum featured a musical laser show. The table at the right shows the number of people who attended each of the shows. Make a frequency table to organize the data.

Attendance					
96	22	57	61	92	80
25	34	76	65	82	37
88	67	42	45	94	64
92	29	33	57	44	88
38	27	75	26	49	91
72	36	89	54	45	90

EXAMPLE

3 The owners of a bookstore specializing in travel books are looking for a new location. They counted the number of people who passed by the proposed location during one afternoon. The frequency table below shows the results of their sampling.

People	Tally	Frequency
under 13	卌	5
teens	卌 |||	8
20s	卌 卌 卌 卌 卌	25
30s	卌 卌 卌 卌 卌 卌 ||	32
40s	卌 卌 卌 | 卌 卌 卌 卌	36
50s	卌 卌 卌 |||	18
60s	卌 卌 |	11

a. Which group of people passed by the location most frequently?

[ ] in their 40s

b. Is this a good location for the bookstore? Explain.

Yes, people in their [ ] are likely to be interested in [ ].

Your Turn **The owners of a children's shoe store are looking for a new location. They counted the number of people who passed by the proposed location during one afternoon. The frequency table below shows the results of their sampling.**

People	Tally	Frequency
under 13	||	2
teens	𝍸𝍸𝍸 |	16
20s	𝍸𝍸 |	11
30s	𝍸𝍸	10
40s	𝍸𝍸𝍸 𝍸𝍸𝍸 ||||	34
50s	𝍸𝍸𝍸 ||||	19
60s	𝍸𝍸 || 𝍸𝍸	22

a. Which group of people passed by the location most frequently?

b. Is this a good location for the shoe store? Explain.

HOMEWORK ASSIGNMENT

Page(s):

Exercises:

1–7 Displaying and Interpreting Data

WHAT YOU'LL LEARN

- Construct and interpret line graphs, histograms, and stem-and-leaf plots.

BUILD YOUR VOCABULARY (page 3)

A **line graph** is one where _____ is displayed to show _____.

FOLDABLES™

ORGANIZE IT

In your notes, list four types of graphs that are used to organize data.

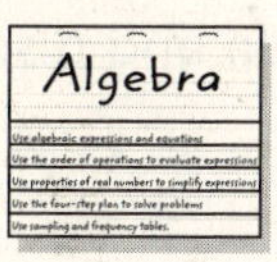

EXAMPLE

1 **Construct a line graph of the data given in the table. Use the graph to predict the percent of the labor force in farming in the year 2010.**

Draw a _____ axis and a vertical axis and label them as shown below. Include the _____. Plot the _____. Draw a _____ by connecting the _____.

Percent of the Labor Force in Farming	
Year	**Percent**
1940	17
1950	12
1960	6
1970	3
1980	2
1990	2

Percent of the Labor Force in Farming

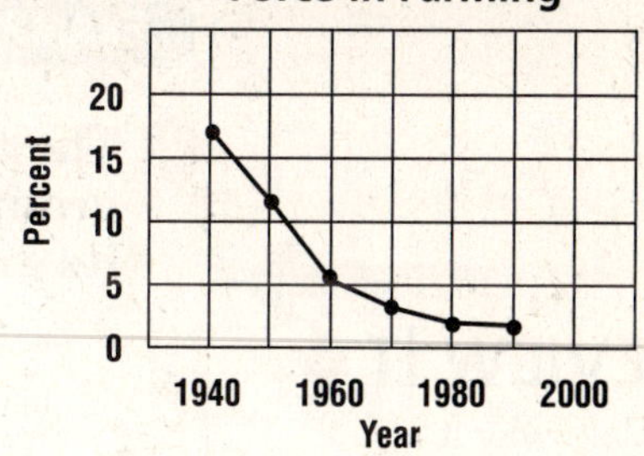

You can see from the graph that the general trend is that the percent of the labor force in farming is _____.

A good prediction for the year 2010 might be between _____.

Your Turn Construct a line graph of the data given in the table. Use the graph to predict the years of life expected at birth for men born in the year 2010.

Life Expectancy for Men, 1910–1990	
Year	**Years of Life**
1910	48.4
1930	58.1
1950	65.6
1970	67.1
1990	71.8

BUILD YOUR VOCABULARY (page 2)

A **histogram** uses data from a frequency table and displays it over ______ intervals.

EXAMPLE

2 The table shows the number of people in different age groups who entered a new store during the first hour of its grand opening. Construct a histogram of the data.

Age	Tally	Frequency
1–10	𝍸	5
11–20	𝍸 |||	8
21–30	𝍸 𝍸 𝍸 𝍸 𝍸	25
31–40	𝍸 𝍸 𝍸 𝍸 𝍸 𝍸 ||	32
41–50	𝍸 𝍸 𝍸 𝍸 𝍸 𝍸 𝍸		36
51–60	𝍸 𝍸 𝍸 |||	18
61–70	𝍸 𝍸		11

REVIEW IT

Explain why it is important to use equal intervals when labeling a histogram.

____

____

____

____

____

____

STEP 1 Draw a horizontal axis and a vertical axis and label them as shown below. Include the title.

STEP 2 Label equal intervals given in the frequency table on the horizontal axis. Label equal intervals of 5 on the ______ axis.

STEP 3 For each time interval, draw a ______ whose height is given by the ______.

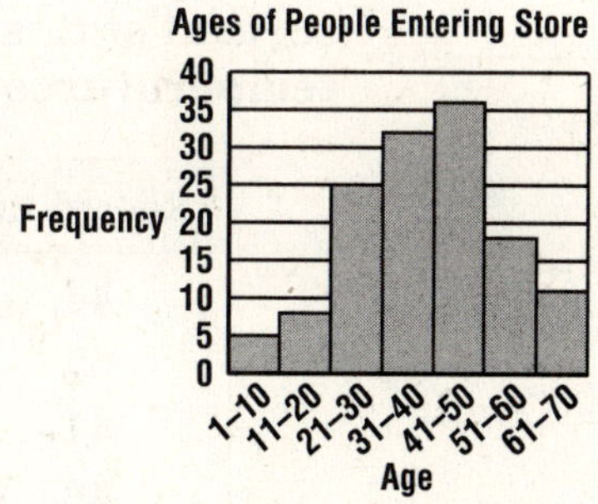

Your Turn The table shows the number of people in different age groups who attended a play on opening night. Construct a histogram of the data.

Age	Tally	Frequency
1–10	||	2
11–20	卌 卌	10
21–30	卌 卌 卌	15
31–40	卌 卌 卌 卌 |||	23
41–50	卌 卌 卌 卌 卌 卌 卌	35
51–60	卌 卌 卌 卌 卌 ||||	29
61–70	卌 卌 ||	12

BUILD YOUR VOCABULARY (page 3)

When data is organized into two ________, stems and leaves, the result is a **stem-and-leaf plot**.

REMEMBER IT

The leaves in a stem-and-leaf plot are always single-digit values.

EXAMPLE

3 The table shows the record high temperatures for several states. Make a stem-and-leaf plot of the temperatures.

Record High Temperatures for Several U.S. States (°C)			
44	38	53	49
57	48	41	43
41	43	45	38
48	47	47	48

The ________ digits are the ________, so the stems are ________. The ________ digits are the ________.

Arrange the leaves in numerical ________ to make the results easier to observe and analyze.

Stem	Leaf
3	________
4	1 1 3 3 4 5 7 7 8 8 8 9
5	________

$3|8 = 38$

Your Turn Make a stem-and-leaf plot of the heights in inches of the following 3rd graders.
48, 46, 39, 38, 45, 46, 49, 42, 50, 52, 42, 48, 47, 44, 51, 43

HOMEWORK ASSIGNMENT

Page(s):
Exercises:

CHAPTER 1

BRINGING IT ALL TOGETHER

STUDY GUIDE

FOLDABLES™	VOCABULARY PUZZLEMAKER	BUILD YOUR VOCABULARY
Use your **Chapter 1 Foldable** to help you study for your chapter test.	To make a crossword puzzle, word search, or jumble puzzle of the vocabulary words in Chapter 1, go to: www.glencoe.com/sec/math/t_resources/free/index.php	You can use your completed **Vocabulary Builder** (pages 2–3) to help you solve the puzzle.

1-1 Writing Expressions and Equations

Write the letter of the algebraic expression that best matches each phrase.

1. three more than a number n
2. five times the difference of x and 4
3. one half the number r
4. the product of x and y divided by 2

a. $5(x - 4)$
b. $\frac{1}{2}r$
c. $n + 3$
d. $\frac{xy}{2}$

5. Translate two times the sum of x and 3 equals 4.

1-2 Order of Operations

Find the value of each expression.

6. $400 - 5[12 + 9]$

7. $69 + 57 \div 3 + 16 \cdot 4$

8. $17 + 3 \cdot 6$

9. $\frac{24 + 3 \cdot 4}{6 \div 2}$

1-3 Commutative and Associative Properties

Write the letter of the term that best matches each equation.

10. $3 + 6 = 6 + 3$

11. $2 + (3 + 4) = (2 + 3) + 4$

12. $2 \cdot (3 \cdot 4) = (2 \cdot 3) \cdot 4$

13. $2 \cdot (3 \cdot 4) = 2 \cdot (4 \cdot 3)$

a. Associative Property of Addition

b. Associative Property of Multiplication

c. Commutative Property of Addition

d. Commutative Property of Multiplication

e. Closure Property

1-4 Distributive Property

14. Tell how you can use the Distributive Property to write $12m + 8m$ in simplest form. Use the word *coefficient* in your explanation.

15. Explain how the Distributive Property could be used to rewrite $5(6 - 4)$.

1-5 A Plan for Problem Solving

16. A 1-ounce serving of chips has 140 calories. There are 14 servings of chips in a bag. How many calories are there in a bag of chips?

1-6

Collecting Data

The principal needs to decide how to select the students to be polled. Determine whether each is a good sample. Explain why or why not.

17. All of the students are asked to enter through the main doors. Every twenty-fifth student is selected to be polled.

18. Only those students who are in the four classrooms closest to the principal's office are selected for the poll.

1-7

Displaying and Interpreting Data

The manager of Fox Music tallied the number of people who applied for jobs at the store.

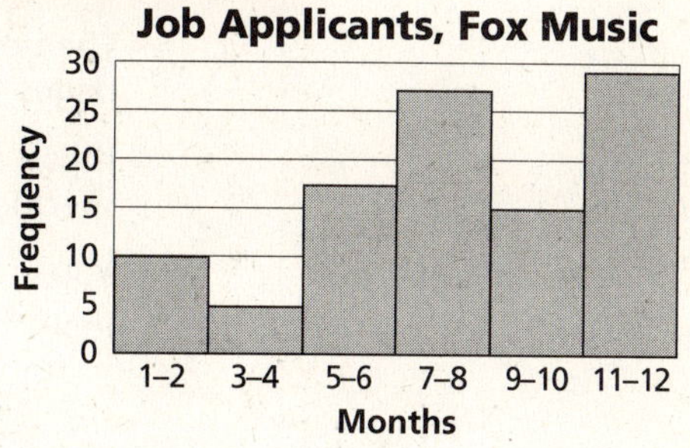

19. In which months was the number of applicants the least?

20. In which months was the number of applicants the most?

Mrs. Anderson's science class recorded the daily high temperature every day for 30 days.

21. Make a stem-and-leaf plot of the data.

High Temperature (°F)					
68	65	68	70	70	70
65	66	68	72	73	76
75	78	81	78	74	71
72	78	77	74	70	68
67	69	68	66	65	63

CHAPTER 1

Checklist

ARE YOU READY FOR THE CHAPTER TEST?

Math Online

Visit **algconcepts.com** to access your textbook, more examples, self-check quizzes, and practice tests to help you study the concepts in Chapter 1.

Check the one that applies. Suggestions to help you study are given with each item.

☐ **I completed the review of all or most lessons without using my notes or asking for help.**

- You are probably ready for the Chapter Test.
- You may want take the Chapter 1 Practice Test on page 47 of your textbook as a final check.

☐ **I used my Foldable or Study Notebook to complete the review of all or most lessons.**

- You should complete the Chapter 1 Study Guide and Review on pages 44–46 of your textbook.
- If you are unsure of any concepts or skills, refer back to the specific lesson(s).
- You may also want to take the Chapter 1 Practice Test on page 47.

☐ **I asked for help from someone else to complete the review of all or most lessons.**

- You should review the examples and concepts in your Study Notebook and Chapter 1 Foldable.
- Then complete the Chapter 1 Study Guide and Review on pages 44–46 of your textbook.
- If you are unsure of any concepts or skills, refer back to the specific lesson(s).
- You may also want to take the Chapter 1 Practice Test on page 47.

Student Signature

Parent/Guardian Signature

Teacher Signature

Integers

FOLDABLES™ Use the instructions below to make a Foldable to help you organize your notes as you study the chapter. You will see Foldable reminders in the margin of this Interactive Study Notebook to help you in taking notes.

Begin with four sheets of plain $8\frac{1}{2}$" × 11" paper.

STEP 1 **Stack**
Stack sheets of paper with edges $\frac{3}{4}$ inch apart.

STEP 2 **Fold**
Fold up bottom edges. All tabs should be the same size.

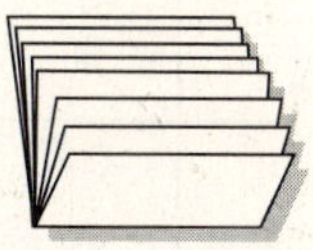

STEP 3 **Staple**
Staple along the fold.

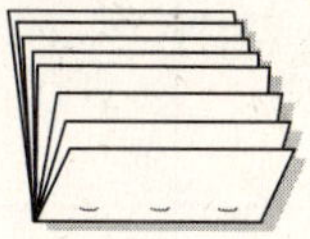

STEP 4 **Label**
Label the tabs as shown.

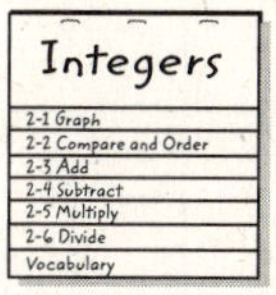

NOTE-TAKING TIP: When taking notes, be sure to focus on what the teacher is saying. Listen for important concepts being emphasized.

CHAPTER 2

BUILD YOUR VOCABULARY

This is an alphabetical list of new vocabulary terms you will learn in Chapter 2. As you complete the study notes for the chapter, you will see Build Your Vocabulary reminders to complete each term's definition or description on these pages. Remember to add the textbook page number in the second column for reference when you study.

Vocabulary Term	Found on Page	Definition	Description or Example
absolute value			
additive inverse [A-duh-tiv]			
coordinate [co-OR-duh-net]			
coordinate plane			
coordinate system			
dimensions			
element			
graph			
integers [IN-tah-jerz]			
matrix [MAY-triks]			
natural numbers			
negative numbers			

Vocabulary Term	Found on Page	Definition	Description or Example
number line			
opposites			
ordered array			
ordered pair			
origin [OR-a-jin]			
quadrants [KWA-druntz]			
scalar multiplication [SKAY-ler]			
Venn diagram			
x-axis			
x-coordinate			
y-axis			
y-coordinate			
zero pair			

2–1 Graphing Integers on a Number Line

What You'll Learn

- Graph integers on a number line and compare and order integers.

Build Your Vocabulary (pages 28–29)

A **number line** is a line with ____ distances marked off to represent numbers.

A **negative number** is a number less than ____.

Diagrams that use ____ or ovals inside a rectangle to show relationships are called **Venn diagrams.**

Natural numbers are the set of whole numbers without the element ____.

To plot points named by numbers on a number line is to **graph**.

The ____ that corresponds to a point on a number line is a **coordinate**.

Key Concept

Integers Integers are the negative numbers −1, −2, −3, −4, . . . and whole numbers 0, 1, 2, 3, 4, . . .

Examples

1 Name the coordinates of *G*, *H*, and *J*.

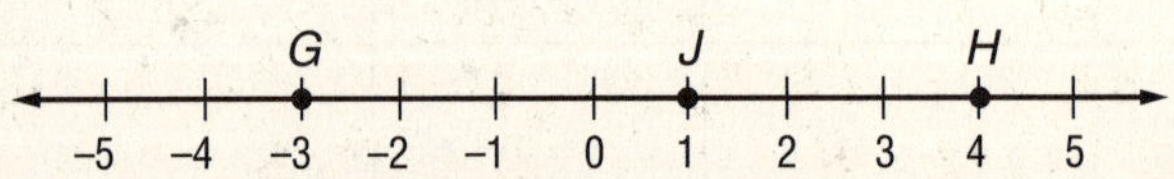

The coordinate of *G* is ____, *H* is ____, and *J* is ____.

2 Graph points *K*, *L*, and *M* on a number line if *K* has coordinate −4, *L* has coordinate 2, and *M* has coordinate −1.

Find each number on a ____. Place a ____ on the mark above the number. Then write the ____ above the dot.

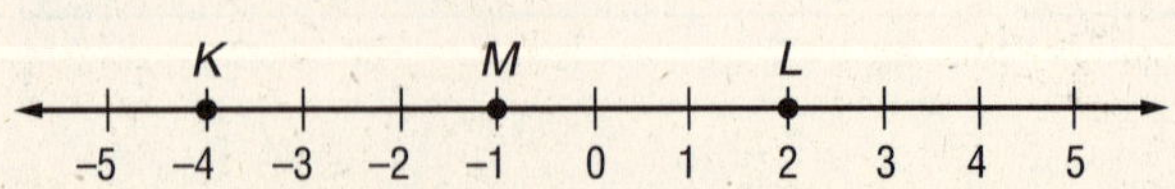

FOLDABLES™

ORGANIZE IT

Copy the Venn diagram from page 52 of your textbook under the tab for Lesson 2-1. Be sure to explain the diagram.

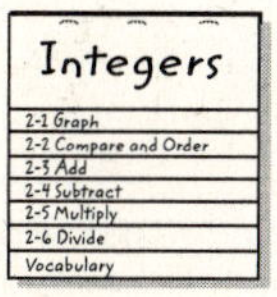

Your Turn

a. Name the coordinates of P, S, and Z.

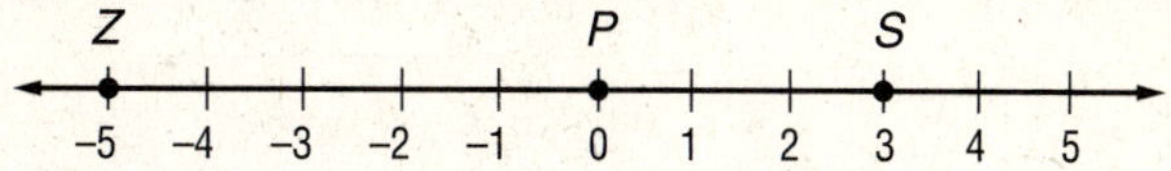

b. Graph points A, B, and C on a number line if A has coordinate -3, B has coordinate -2, and C has coordinate 5.

EXAMPLE

3 Replace ● in 3 ● −4 with < or > to make a true sentence.

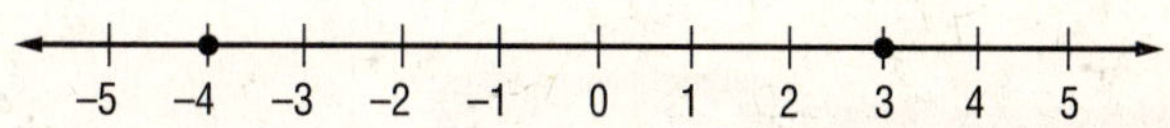

3 is to the ______ of -4 on the number line. So, 3 ______ -4.

EXAMPLE

WRITE IT

Write an example of a negative number in a real world application.

______

______

______

______

4 The table shows the high temperature each day for one week in January in a midwestern city. Order the temperatures from greatest to least.

Day	Temp. (°C)
Monday	5
Tuesday	−1
Wednesday	8
Thursday	−3
Friday	0
Saturday	−7
Sunday	−2

Graph each ______ on a number line.

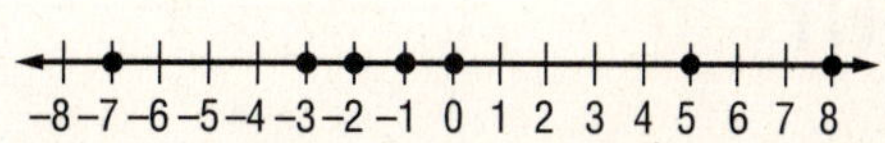

The order from greatest to least is ______, 5°, ______, ______, −2°, ______, and −7°.

Your Turn **Replace ● with < or > to make a true sentence.**

a. −4 ● 2

b. 5 ● −6

c. The table shows the average low temperature each day for one week in January in a northern city. Order the temperatures from least to greatest.

City	Temp. (°C)
Monday	7
Tuesday	5
Wednesday	−3
Thursday	−5
Friday	−1
Saturday	0
Sunday	4

KEY CONCEPT

Absolute Value The absolute value of a number is the distance it is from 0 on the number line.

EXAMPLES

5 **Evaluate each expression.**

$|-1|$

$|-1| =$ ____ The graph of −1 is ____ unit away from ____.

6 $|4| + |-5|$

$|4| + |-5| =$ ____ + ____ The absolute value of 4 is ____.

$= 9$ The absolute value of −5 is ____.

Your Turn **Evaluate each expression.**

a. $|-4|$

b. $|-3| + |7|$

HOMEWORK ASSIGNMENT

Page(s):

Exercises:

2–2 The Coordinate Plane

What You'll Learn

- Graph points on a coordinate plane.

Build Your Vocabulary (pages 28–29)

The **coordinate system** is the ______ formed by the intersection of two perpendicular number lines that meet at their zero points.

The point of intersection of the two ______ in the coordinate plane is called the **origin**.

The ______ number line on a coordinate plane is the ***y*-axis**. The horizontal number line is the ***x*-axis**.

The plane containing the *x*- and *y*-axes is the **coordinate plane**.

An **ordered pair** is a pair of numbers used to locate any ______ on a ______.

The ______ number in a coordinate pair is the ***x*-coordinate**.

The ______ number in a coordinate pair is the ***y*-coordinate**.

EXAMPLE

1 Write the ordered pair that names point *H*.

The x-coordinate is ______,

and the y-coordinate is ______.

The ordered pair for point H is ______.

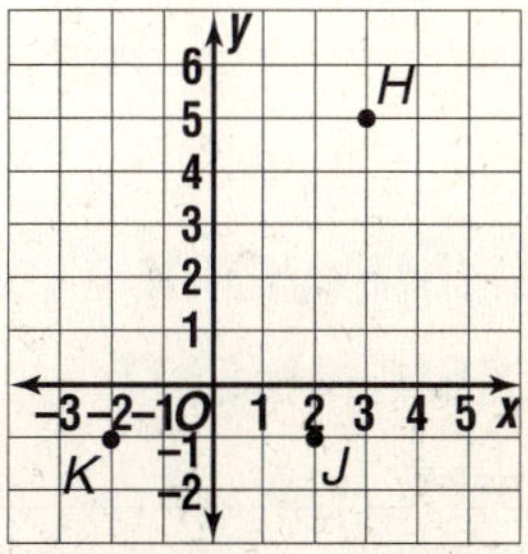

Your Turn **Write the ordered pair that names each point.**

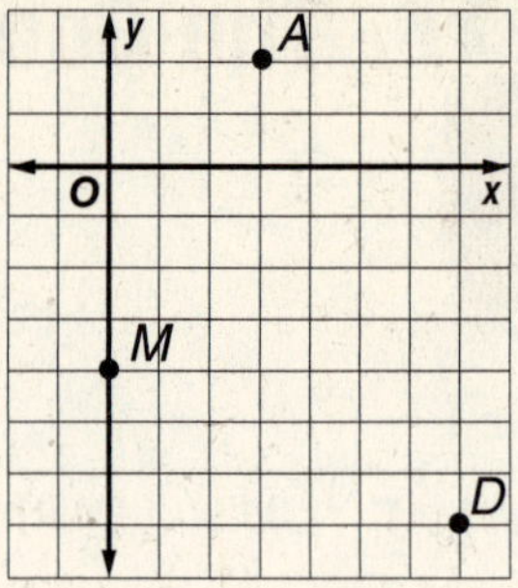

a. *A* ______

b. *M* ______

c. *D* ______

EXAMPLE

2 **Graph *V*(2, 4) on the coordinate plane.**

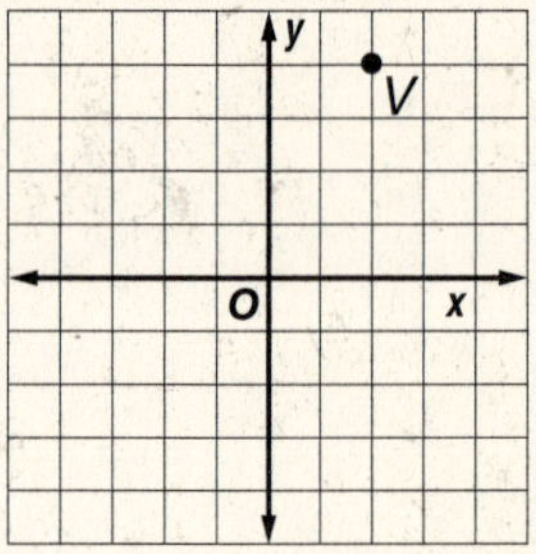

- Start at the ______, *O*.
- The *x*-coordinate is ______. So, move 2 units to the ______.
- The *y*-coordinate is ______. So, move 4 units ______ and draw a ______.

FOLDABLES

ORGANIZE IT

Under the tab for Lesson 2-2, draw a coordinate plane and label each of the four quadrants as well as the *x*- and *y*-axes.

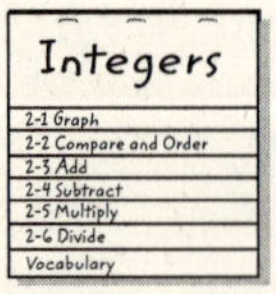

Your Turn **Graph each point on a coordinate plane.**

a. *B*(5, 0)

b. *G*(−5, 3)

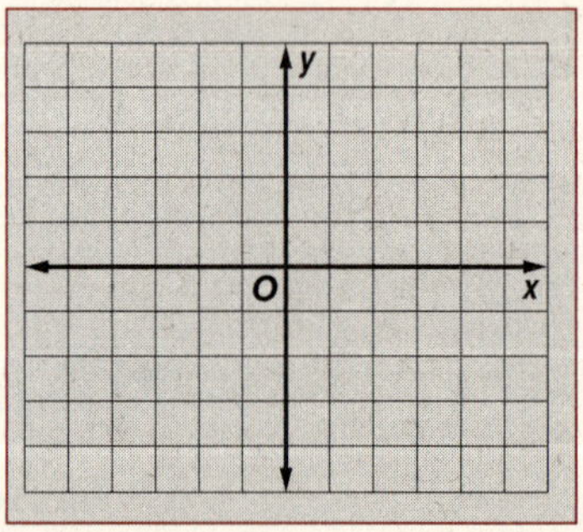

EXAMPLE

3 **Name the quadrant in which *F*(0, 1) is located.**

Point *F* lies on the ______. It is not located in a ______.

Your Turn **Name the quadrant in which each point is located.**

a. *S*(−2, −4) ______

b. *H*(−8, 1) ______

HOMEWORK ASSIGNMENT

Page(s):

Exercises:

2–3 Adding Integers

WHAT YOU'LL LEARN

- Add integers.

KEY CONCEPTS

Adding Integers with the Same Sign To add integers with the same sign, add their absolute values. Give the result the same sign as the integers.

Additive Inverse Property The sum of any number and its additive inverse is 0.

Adding Integers with Different Signs To add integers with different signs, find the difference of the absolute values. Give the result the same sign as the integer with the greater absolute value.

EXAMPLES

Find each sum.

1 **$6 + 7$**

$6 + 7 =$ ______ Both numbers are ______, so the sum is ______.

2 **$-5 + (-8)$**

$-5 + (-8) =$ ______ Both numbers are ______, so the sum is ______.

Your Turn **Find each sum.**

a. $7 + 15$

______

b. $-14 + (-17)$

______

BUILD YOUR VOCABULARY (pages 28–29)

When one ______ algebra tile is paired with one ______ algebra tile, the result is a **zero pair**.

If the ______ of two numbers is ______, the numbers are called **opposites** or **additive inverses**.

EXAMPLES

Find each sum.

3 **$-9 + 5$**

$|-9| - |5| =$ ______ $-$ ______ or ______

$|-9| > |5|$, so the sum is ______.

Therefore, $-9 + 5 =$ ______.

FOLDABLES™

ORGANIZE IT

Write an expression that is the sum of one negative integer and one positive integer under the tab for Lesson 2-3. Tell how you know whether the sum is positive or negative.

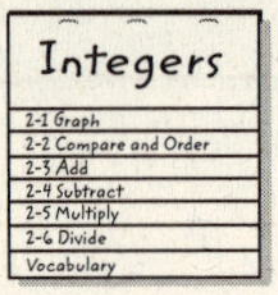

4 **$(-9) + 8$**

$|-9| - |8| =$ ____ $-$ ____ or ____.

$|-9| > |8|$, so the sum is ____.

Therefore, $(-9) + 8 =$ ____.

Your Turn **Find each sum.**

a. $-5 + 2$

b. $(-1) + 9$

EXAMPLE

5 **Simplify $-7y + 6y$.**

$-7y + 6y = [$ ____ $+ 6]y$ Use the ____ Property.

$=$ ____ $-7 + 6 =$ ____

Your Turn Simplify $3a + (-5a)$.

HOMEWORK ASSIGNMENT

Page(s):

Exercises:

2–4 Subtracting Integers

What You'll Learn

- Subtract integers.

Key Concept

Subtracting Integers To subtract an integer, add its additive inverse.

Foldables™

Organize It

Write an expression that is the difference of two negative integers under the tab for Lesson 2-4. Rewrite the expression as an addition expression.

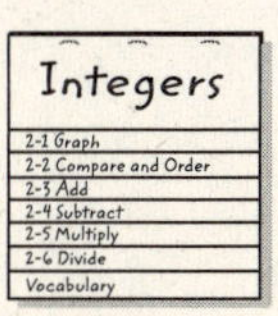

EXAMPLES

Find each difference.

1. **10 − 3**

$10 - 3 = 10 + $ ___ To subtract 3, add ___.

$=$ ___

2. **−7 − (−6)**

$-7 - (-6) = -7 + $ ___ To subtract −6, add ___.

$=$ ___

3. **−1 − 8**

$-1 - 8 = -1 + $ ___ To subtract 8, add ___.

$=$ ___

4. **3 − (−5)**

$3 - (-5) = 3 + $ ___ To subtract −5, add ___.

$=$ ___

5. **4 − 6**

$4 - 6 = 4 + $ ___ To subtract 6, add ___.

$=$ ___

6. **−7 − (−10)**

$-7 - (-10) = -7 + $ ___ To subtract −10, add ___.

$=$ ___

Your Turn **Find each difference.**

a. $9 - 6$ **b.** $-8 - (-2)$ **c.** $-4 - 6$

d. $7 - (-2)$ **e.** $1 - 8$ **f.** $-2 - (-5)$

EXAMPLES

7 Evaluate $a - b$ if $a = -8$ and $b = -2$.

$a - b = -8 - (-2)$ $a = -8, b = -2.$

$=$ ______ Write $-8 - (-2)$ as ______.

$=$ ______ $-8 + 2 =$ ______

8 Evaluate $p + q - r$ if $p = -7$, $q = -3$, and $r = 2$.

$p + q - r = -7 + (-3) - 2$ $p = -7, q = -3$, and $r = 2$.

$=$ ______ $- 2$ $-7 + (-3) =$ ______

$=$ ______ Write $-10 - 2$ as ______.

$=$ ______ $-10 + (-2) =$ ______

Your Turn

a. Evaluate $m - n$ if $m = -5$ and $n = -4$.

b. Evaluate $a + b - c$ if $a = -3$, $b = -8$, and $c = 5$.

HOMEWORK ASSIGNMENT

Page(s):

Exercises:

2–5 Multiplying Integers

WHAT YOU'LL LEARN

- Multiply integers.

KEY CONCEPTS

Multiplying Two Integers with Different Signs
The product of two integers with different signs is negative.

Multiplying Two Integers with the Same Signs
The product of two integers with the same signs is positive.

EXAMPLES

Find each product.

1 **4(−3)**

$4(-3) =$ ______ The factors have ______ signs.

The product is ______.

2 **−2(7)**

$-2(7) =$ ______

Your Turn **Find each product.**

a. $2(-2)$

b. $-3(10)$

EXAMPLES

Find each product.

3 **10(4)**

$10(4) =$ ______ The factors have the ______ sign.

The product is ______.

4 **−8(−6)**

$-8(-6) =$ ______

Your Turn **Find each product.**

a. $5(3)$

b. $-1(-8)$

EXAMPLE

5 Find 7(−3)(−6).

$7(-3)(-6) = $ ____ (-6) — $7(-3) = $ ____

$= $ ____ — $-21(-6) = $ ____

Your Turn **Find each product.**

a. $3(-4)(-2)$

____

b. $2(-1)(-9)(5)$

____

EXAMPLES

6 Evaluate $4ab$ if $a = -3$ and $b = -5$.

$4ab = 4(-3)(-5)$ — Replace a with -3 and b with -5.

$= $ ____ (-5) — $4(-3) = $ ____

$= $ ____ — $-12(-5) = $ ____

7 Simplify $(4m)(-7n)$.

$(4m)(-7n) = (4)$ ____ (-7) ____ — $4m = (4)$ ____; $-7n = (-7)$ ____

$= (4)(-7)$ ____ — Commutative Property

$= $ ____ — $(4)(-7) = $ ____; $(m)(n) = $ ____

FOLDABLES™

ORGANIZE IT

Write a multiplication expression using two negative integers under the tab for Lesson 2-5. Will the solution be positive or negative? How do you know?

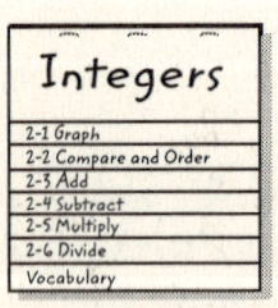

Your Turn

a. Evaluate $3xy$ if $x = -7$ and $y = -3$.

____

b. Simplify $(-3n)(2x)$.

____

HOMEWORK ASSIGNMENT

Page(s):

Exercises:

2–6 Dividing Integers

WHAT YOU'LL LEARN

- Divide integers.

KEY CONCEPT

Dividing Integers The quotient of two integers with the same sign is positive. The quotient of two integers with different signs is negative.

EXAMPLES

Find each quotient.

1 $\mathbf{-12 \div 3}$

$-12 \div 3 =$ ______ The signs are ______.

The quotient is ______.

2 $\mathbf{-50 \div (-10)}$

$-50 \div (-10) =$ ______ The signs are the ______.

The quotient is ______.

Your Turn **Find each quotient.**

a. $-10 \div 5$

b. $-35 \div (-7)$

EXAMPLE

3 **Evaluate $\frac{3a}{b}$ if $a = -6$ and $b = 9$.**

$\frac{3a}{b} = \frac{3(-6)}{9}$ Replace a with ______ and b with ______.

$= \frac{\text{______}}{9}$ $3(-6) =$ ______

$=$ ______ Divide ______ by ______.

Your Turn Evaluate $\frac{5x}{y}$ if $x = -4$ and $y = 2$.

EXAMPLE

FOLDABLES™

ORGANIZE IT

In your own words, summarize the rules for dividing integers under the tab for Lesson 2-6.

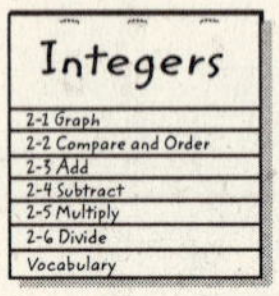

4 In the last 5 years at a high school, the number of students with no tardies during the entire school year dropped from 315 to 95. What was the average change in the number of students without a tardy for each of those 5 years?

First, find the change in the number of students with no tardies.

____ − ____ = −220 There were ____ fewer students with no tardies at the end of the ____ years.

To find the average change, divide −220 by ____.

−220 ÷ ____ = ____

The ____ change in the number of students with no tardies for the entire school year was ____ per ____.

Your Turn In 1990, the population of Washington, D.C., was 606,900. By 1998, the population of the city had fallen to 523,124. Find the average change in population for each of those eight years.

HOMEWORK ASSIGNMENT

Page(s):

Exercises:

CHAPTER 2

BRINGING IT ALL TOGETHER

STUDY GUIDE

FOLDABLES™	VOCABULARY PUZZLEMAKER	BUILD YOUR VOCABULARY
Use your **Chapter 2 Foldable** to help you study for your chapter test.	To make a crossword puzzle, word search, or jumble puzzle of the vocabulary words in Chapter 2, go to: www.glencoe.com/sec/math/t_resources/free/index.php	You can use your completed **Vocabulary Builder** (pages 28–29) to help you solve the puzzle.

2-1

Graphing Integers on a Number Line

Refer to the Venn diagram shown. Write *true* or *false* for each of the following statements.

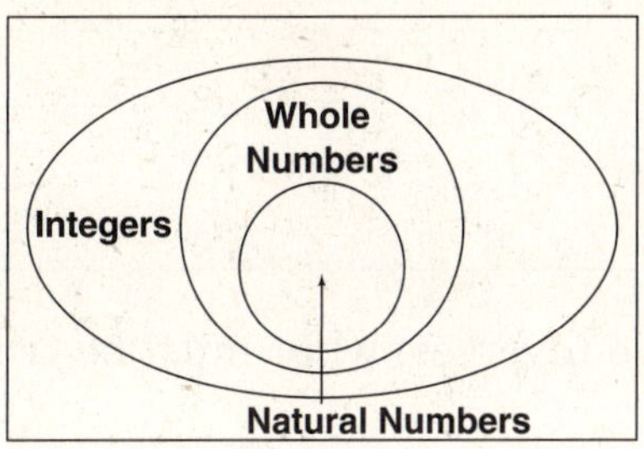

1. All whole numbers are integers.

2. All natural numbers are integers.

3. All whole numbers are natural numbers.

4. All natural numbers are whole numbers.

Replace each ● with $<$ or $>$ to make a true sentence.

5. $6 \bullet -3$ **6.** $4 \bullet -5$ **7.** $|-3| \bullet |2|$ **8.** $|-1| \bullet |-8|$

2-2

The Coordinate Plane

Graph the given points on the coordinate plane shown.

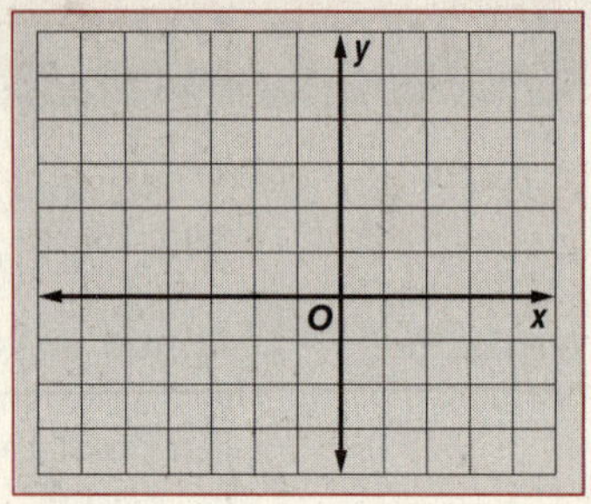

9. $A(0, 3)$

10. $B(1, 4)$

11. $C(-4, 2)$

12. $D(-6, -1)$

2-3

Adding Integers

13. Explain how to add integers with the same sign.

14. Explain how to add integers with opposite signs.

15. If two numbers are additive inverses, what must be true about their absolute values?

Find each sum.

16. $-3 + (-6)$

17. $4 + (-1)$

18. $7 + (-10)$

19. $-1 + 10$

2-4 Subtracting Integers

Write each subtraction problem as an addition problem.

20. 12 − 4

21. −15 − 7

22. 0 − 11

23. −15 − (−4)

Describe how to find each difference. Then find each difference.

24. 8 − 11

25. 5 − (−8)

26. 17 − 14

27. −8 − 19

2-5 Multiplying Integers

Find each product.

28. (−4)(9)

29. (−2)(−13)

30. 5(−8)

31. 6(3)

32. 4(−3)

33. −7(−10)

Find each quotient.

34. 15 ÷ 12

35. 9 ÷ −1

36. $\frac{35}{-7}$

37. $\frac{-78}{-13}$

CHAPTER 2

Checklist

ARE YOU READY FOR THE CHAPTER TEST?

Math Online

Visit **algconcepts.com** to access your textbook, more examples, self-check quizzes, and practice tests to help you study the concepts in Chapter 2.

Check the one that applies. Suggestions to help you study are given with each item.

☐ **I completed the review of all or most lessons without using my notes or asking for help.**

- You are probably ready for the Chapter Test.
- You may want take the Chapter 2 Practice Test on page 89 of your textbook as a final check.

☐ **I used my Foldable or Study Notebook to complete the review of all or most lessons.**

- You should complete the Chapter 2 Study Guide and Review on pages 86–88 of your textbook.
- If you are unsure of any concepts or skills, refer back to the specific lesson(s).
- You may also want to take the Chapter 2 Practice Test on page 89.

☐ **I asked for help from someone else to complete the review of all or most lessons.**

- You should review the examples and concepts in your Study Notebook and Chapter 2 Foldable.
- Then complete the Chapter 2 Study Guide and Review on pages 86–88 of your textbook.
- If you are unsure of any concepts or skills, refer back to the specific lesson(s).
- You may also want to take the Chapter 2 Practice Test on page 89.

Student Signature

Parent/Guardian Signature

Teacher Signature

CHAPTER 3

Addition and Subtraction Equations

FOLDABLES™ Use the instructions below to make a Foldable to help you organize your notes as you study the chapter. You will see Foldable reminders in the margin of this Interactive Study Notebook to help you in taking notes.

Begin with a sheet of 11" × 17" paper.

STEP 1 **Fold**
Fold the short sides to meet in the middle.

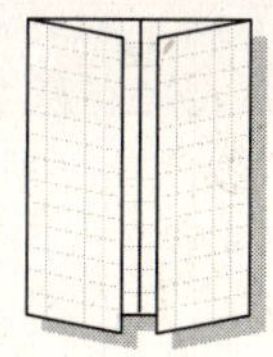

STEP 2 **Fold**
Fold the top to the bottom.

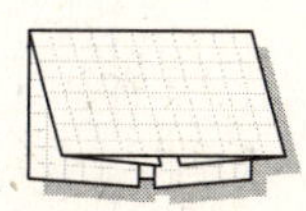

STEP 3 **Open**
Cut along second fold to make four tabs.

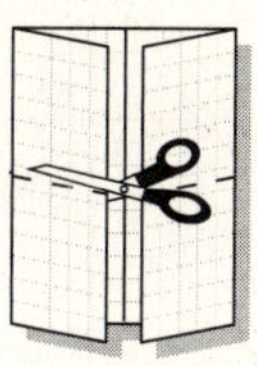

STEP 4 **Label**
Label each tab as shown.

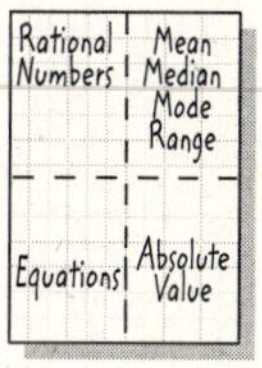

NOTE-TAKING TIP: It is often helpful to review your notes as soon as you can after class.

CHAPTER 3

BUILD YOUR VOCABULARY

This is an alphabetical list of new vocabulary terms you will learn in Chapter 3. As you complete the study notes for the chapter, you will see Build Your Vocabulary reminders to complete each term's definition or description on these pages. Remember to add the textbook page number in the second column for reference when you study.

Vocabulary Term	Found on Page	Definition	Description or Example
cross products			
empty set			
equivalent equations			
inequality [IN-ee-KWAL-a-tee]			
mean			
measure of central tendency			
measure of variation			
median			

Vocabulary Term	Found on Page	Definition	Description or Example
mode			
open sentence			
range			
rational numbers [RA-shun-ul]			
replacement set			
sequence [SEE-kwens]			
solution			
solving			
statement			
unit cost			

3–1 Rational Numbers

BUILD YOUR VOCABULARY (page 48)

An **inequality** is a mathematical sentence that uses ______ and ______ to compare two expressions.

Cross products are the products of the ______ terms of two ______.

WHAT YOU'LL LEARN

- Compare and order rational numbers.

KEY CONCEPTS

Rational Number A rational number is any number that can be expressed as a fraction where the numerator and denominator are integers and the denominator is not zero.

Comparison Property For any two numbers *a* and *b*, exactly one of the following sentences is true.
$a < b \quad a > b \quad a = b$

EXAMPLES

Replace each ● with <, >, or = to make a true sentence.

1 -1 ● $-\frac{3}{5}$

Since -1 is to the ______ of $-\frac{3}{5}$ on a number line,

-1 ______ $-\frac{3}{5}$.

2 $-3(2)(0)$ ● $7 + (-8)$

$-3(2)(0)$ ● $7 + (-8)$

______ ● ______ Find the value of each side.

0 is to the ______ of -1 on a number line and 0 ______ -1.

So, $-3(2)(0)$ ______ $7 + (-8)$.

Your Turn **Replace each ● with <, >, or = to make a true sentence.**

a. -2 ● $\frac{1}{3}$

b. $7(0)$ ● $9 + (-9)$

FOLDABLES™

ORGANIZE IT

Write three examples of rational numbers under the tab for Rational Numbers. Include a decimal and one negative number.

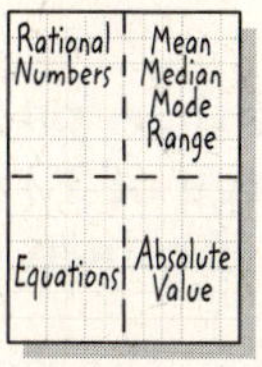

EXAMPLES

Replace each ● with <, >, or = to make each sentence true.

3 $\frac{7}{12}$ ● $\frac{5}{8}$

$\frac{7}{12}$ ● $\frac{5}{8}$

____ (7) ● 12 ____ Find the cross products.

____ < 60

So, $\frac{7}{12}$ ____ $\frac{5}{8}$.

4 $-\frac{4}{15}$ ● $-\frac{3}{10}$

$\frac{-4}{15}$ ● $\frac{-3}{10}$

____ (−4) ● 15 ____

$-40 > -45$

So, $-\frac{4}{15}$ ____ $-\frac{3}{10}$.

Your Turn **Replace each ● with <, >, or = to make each sentence true.**

a. $\frac{2}{3}$ ● $\frac{5}{7}$

b. $-\frac{9}{11}$ ● $-\frac{7}{8}$

EXAMPLE

5 **Write $\frac{5}{6}$, $\frac{7}{9}$, and $\frac{4}{5}$ in order from least to greatest.**

$\frac{5}{6} = 0.8333\ldots$ or ____ This is a ____ decimal.

$\frac{7}{9} = 0.7777\ldots$ or ____ This is a ____ decimal.

$\frac{4}{5} =$ ____ This is a ____ decimal.

In order from least to greatest, the decimals are ____. So, the fractions in order from least to greatest are ____.

Your Turn Write $\frac{2}{3}$, $\frac{5}{8}$, and $\frac{3}{5}$ in order from least to greatest.

BUILD YOUR VOCABULARY (page 49)

Unit cost is the cost per unit.

unit cost = total ______ ÷ number of ______

EXAMPLE

6 Latisha needs to buy snacks for her art club. A package of 12 granola bars costs $2.69 and a package of 18 granola bars costs $3.55. Which is the better buy? Explain.

Find the unit cost of each package. In each case, the unit cost is expressed in cents per bar.

unit cost of package of 12:

$\$2.69 \div$ ______ ≈ 0.2241 or about ______ per bar.

unit cost of package of 18:

$\$3.55 \div$ ______ ≈ 0.1972 or about ______ per bar.

Since $0.20 ______ $0.22, the package of ______ granola bars is the better buy.

Your Turn John needs to buy cookies. The cost of a package of 24 cookies is $3.79. A package of 36 cookies costs $4.29. Which is the better buy? Explain.

HOMEWORK ASSIGNMENT

Page(s):

Exercises:

3–2 Adding and Subtracting Rational Numbers

WHAT YOU'LL LEARN

- Add and subtract rational numbers.

EXAMPLES

Find each sum.

1 $\mathbf{0.41 + (-1.3)}$

$0.41 + (-1.3)$

$= -(|\_\_\_| - |\_\_\_|)$ The signs differ. Subtract the lesser absolute value from the greater.

$= -(\_\_\_ - \_\_\_)$

$= \_\_\_$

REVIEW IT

Explain the procedure for finding the LCD of two fractions. (previous course)

____

____

____

____

2 $\mathbf{-5\frac{3}{10} + 2\frac{3}{5}}$

$-5\frac{3}{10} + 2\frac{3}{5}$

$= \left(-5\frac{3}{10}\right) + 2\_\_\_$ The LCD is 10. Replace $\frac{3}{5}$ with ___.

$= -\left(\left|\_\_\_\right| - \left|\_\_\_\right|\right)$ The signs differ. Subtract the lesser absolute value from the greater.

$= -\left(\_\_\_ - \_\_\_\right)$

$= \_\_\_$

3 $\mathbf{-3.7 + 12.5 + (-1.3)}$

$-3.7 + 12.5 + (-1.3)$

$= \_\_\_ + \left(\_\_\_ + (-1.3)\right)$ Commutative & Associative Properties (+)

$= \_\_\_ + \left(\_\_\_\right)$ Add.

$= \_\_\_$

Your Turn **Find each sum.**

a. $(-3.2) + (-1.1)$

b. $-2\frac{3}{4} + 5\frac{5}{8}$

c. $-20.5 + 10.8 + (-5.4)$

EXAMPLE

4 Find −4.7 − (−5.9)

$-4.7 - (-5.9)$

$= -4.7 +$ ☐ To subtract −5.9 add ☐.

$=$ ☐ The signs differ. Subtract the lesser absolute value from the greater.

Your Turn Find $-8.1 - (-2.4)$.

EXAMPLE

5 Evaluate $c - d$ if $c = 5\frac{1}{6}$ and $d = -3\frac{5}{8}$.

$c - d = 5\frac{1}{6} - \left(-3\frac{5}{8}\right)$ $c = 5\frac{1}{6}$, $d = -3\frac{5}{8}$

$= 5\frac{1}{6} + 3\frac{5}{8}$ To subtract $-3\frac{5}{8}$, add $3\frac{5}{8}$.

$= 5$ ☐ $+ 3$ ☐ The LCD is ☐.

$=$ ☐ $+$ ☐

$=$ ☐ or ☐

HOMEWORK ASSIGNMENT

Page(s):

Exercises:

Your Turn Evaluate $x - y$ if $x = 8\frac{1}{4}$ and $y = -4\frac{2}{5}$.

3–3 Mean, Median, Mode, and Range

What You'll Learn

- Find the mean, median, mode, and range of a set of data.

BUILD YOUR VOCABULARY (page 48)

A **measure of central tendency** is a number used to describe a set of data because it represents a centralized, or ________, value.

Key Concepts

Mean The mean, or *average*, of a set of data is the sum of the data divided by the number of pieces of data.

Median The median of a set of data is the middle number when the data in the set are arranged in numerical order.

Mode The mode of a set of data is the number that occurs most often in the set.

EXAMPLES

The stem-and-leaf plot shows the number of children enrolled in each of 9 gymnastics classes offered at a local recreation center.

Stem	Leaf
0	4 5 8
1	0 2 5 5 8
2	1 *2∣1 = 21*

1 Find the mean of the gymnastics data.

First, find the ________ of the number of children enrolled.

Then ________ by the number of items of data.

$$\text{mean} = \frac{4 + 5 + 8 + 10 + 12 + 15 + 15 + 18 + 21}{\square}$$

$$= \frac{\square}{9} \text{ or } \square$$

2 Find the median of the gymnastics data.

Arrange the numbers in order from ________ to ________.

4 ____ 8 10 ____ 15 15 18 ____

Since there is an ________ number of data items, the ________ number is the median.

The median is ________ students.

3 **Find the mode of the gymnastics data.**

Look for the number that occurs most often.

4 5 8 10 12 15 15 18 21

In this set ______ appears ______. So, the set of data has one mode, ______ students.

Your Turn The stem-and-leaf plot below shows the number of people enrolled in each of 9 aerobics classes offered at a local recreation center. Find the mean, median, and mode of the aerobics data.

Stem	Leaf
0	5 7 8
1	0 3 3 5 6
2	1 *2∣1 = 21*

BUILD YOUR VOCABULARY (page 48)

Measures of variation are used to describe the ______ of the data.

KEY CONCEPT

Range The range of a set of data is the difference between the greatest and the least values of the set.

FOLDABLES™ Include mean, median, mode, and range in your notes. Be sure to include an example of each.

EXAMPLE

4 **Using the stem-and-leaf plot from Examples 1 through 3, find the range of the gymnastics data.**

To find the range of the gymnastics data, subtract the ______ value of the data from the ______ value.

The greatest value is ______.

The least value is ______.

So, the range is ______ – ______ or ______ students.

Your Turn Refer to Example 1. Find the range of the aerobics data.

HOMEWORK ASSIGNMENT

Page(s):

Exercises:

3–4

Equations

WHAT YOU'LL LEARN

- Determine whether a given number is a solution of an equation.

BUILD YOUR VOCABULARY (page 49)

A **statement** is any sentence that is ______ or ______, but not both.

A mathematical sentence like $m - 5 = 12$ is called an **open sentence**.

A set of numbers from which replacements for a ______ may be chosen is called a **replacement set**.

Finding the replacements for the variable that results in a ______ sentence is called **solving**.

The **solutions** of an open sentence are ______ values that make the sentence true.

EXAMPLE

1 Find the solution of $13 = 33 + 4d$ if the replacement set is $\{-6, -5, -4, -3\}$.

REMEMBER IT

A replacement set contains numbers that may result in a false sentence.

Value of d	$13 = 33 + 4d$	True or False?
-6	$13 \stackrel{?}{=} 33 +$ ______	______
-5	$13 \stackrel{?}{=} 33 +$ ______	______
-4	$13 \stackrel{?}{=} 33 +$ ______	______
-3	$13 \stackrel{?}{=} 33 +$ ______	______

Since ______ makes the sentence $13 = 33 + 4d$ true, the solution is ______.

Your Turn **Find the solution of each equation if the replacement set for x is {3, 4, 5, 6} and for n is {0, 1, 2, 3}.**

a. $4x - 7 = 13$

b. $n - 2 = \frac{n-1}{2}$

EXAMPLES

Solve each equation.

2 $h = [24 \div (-3)(2)] - 1$

$h = [24 \div (-3)(2)] - 1$

$h = (24 \div \_\_\_) - 1$ — ___ −3 and 2.

$h = \_\_\_ - 1$ — Divide 24 by ___.

$h = \_\_\_$ — ___ 1 from ___.

The solution is ___.

3 $\frac{4 \cdot 5 - 6}{(3 \cdot 4) - 5} = c$

$\frac{4 \cdot 5 - 6}{(3 \cdot 4) - 5} = c$ — Evaluate the numerator and the denominator separately.

$\frac{\_\_\_}{\_\_\_} = c$ — $4 \cdot 5 - 6 =$ ___ $- 6$ or ___

___ $= c$ — $(3 \cdot 4) - 5 =$ ___ $- 5$ or ___

The solution is ___.

REVIEW IT

Explain the difference between an expression and an equation. *(Lesson 1-1)*

HOMEWORK ASSIGNMENT

Page(s):

Exercises:

Your Turn **Solve each equation.**

a. $x = [20 \div (-4)(1)] - 3$

b. $3(7) - 4 = z$

3–5 Solving Equations By Using Models

WHAT YOU'LL LEARN

- Solve addition and subtraction equations by using models.

EXAMPLES

Use algebra tiles to solve each equation.

1 $t + 3 = 1$

STEP 1 Model $t + 3 = 1$.

Place ______ variable tile and ______ positive square tiles on one side of the mat to represent ______. Place ______ positive square tile on the other side of the mat to represent ______.

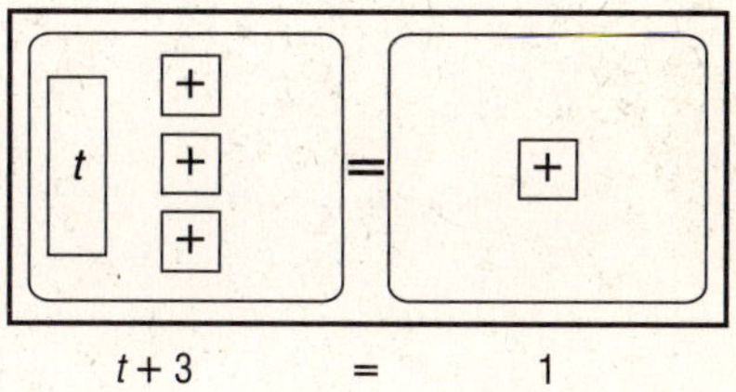

STEP 2 To get the variable tile by itself, add ______ negative square tiles to each side. On the left side you have ______ zero pairs, and on the right you have ______ zero pair.

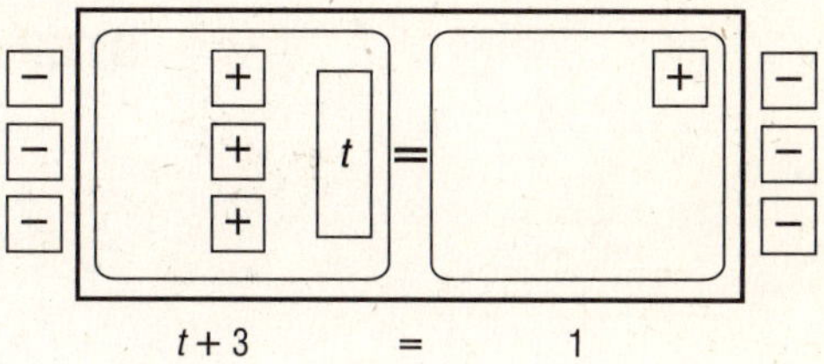

STEP 3 Remove the ______.

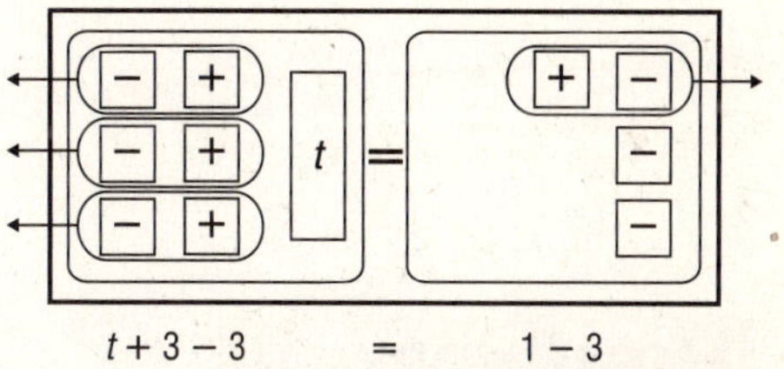

STEP 4 The variable tile is matched with ______ negative square tiles.

So, ______.

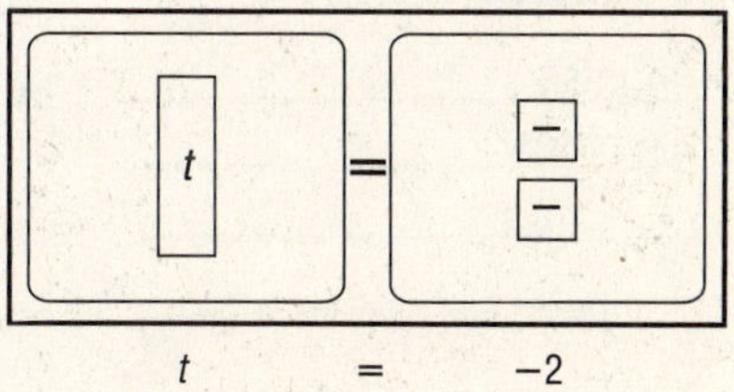

2 $b - 4 = -3$

STEP 1 Write the equation as $b +$ ______ $= -3$. Place ______ variable tile and ______ negative square tiles on one side of the mat to represent ______. Place ______ negative square tiles on the other side of the mat to represent ______.

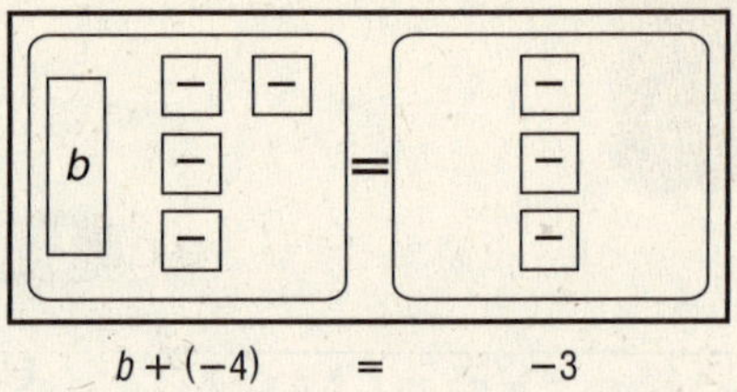

STEP 2 To get the variable tile by itself, add ______ square positive tiles to each side.

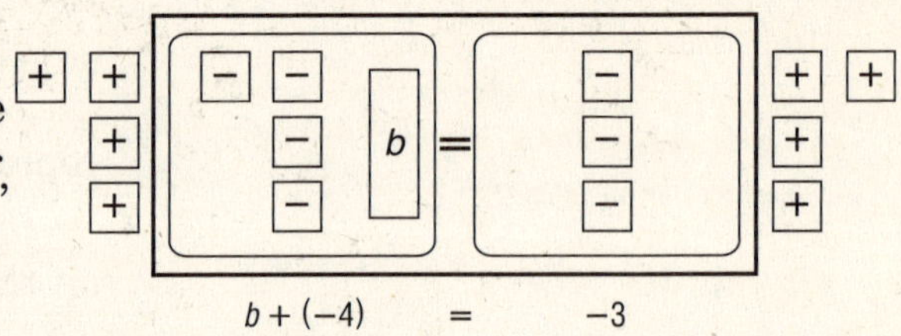

STEP 3 Remove the ______.

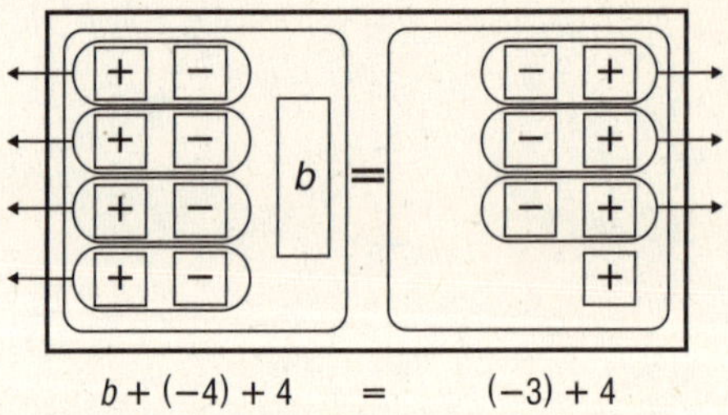

STEP 4 The variable tile is matched with ______ positive square tile.

So, ______

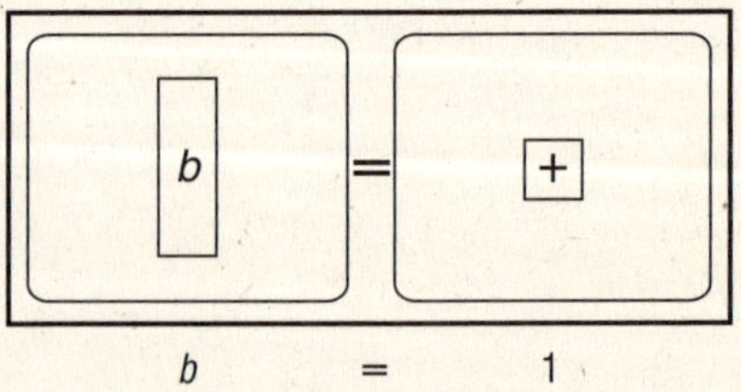

HOMEWORK ASSIGNMENT

Page(s):

Exercises:

Your Turn **Use algebra tiles to solve each equation.**

a. $x - 4 = 6$

b. $p + 4 = 2$

3–6 Solving Addition and Subtraction Equations

WHAT YOU'LL LEARN

- Solve addition and subtraction equations by using the properties of equality.

BUILD YOUR VOCABULARY (page 48)

When the same number is added to each side of an equation, the result is an **equivalent equation**.

KEY CONCEPTS

Addition Property of Equality If you add the same number to each side of an equation, the two sides remain equal.

Subtraction Property of Equality If you subtract the same number from each side of an equation, the two sides remain equal.

FOLDABLES Write the properties under the tab for Equations.

EXAMPLES

Solve each equation.

1 $r + (-13) = 15$

$r + (-13) = 15$

$r - 13 = 15$ — Rewrite the equation.

$r - 13 + \square = 15 + \square$ — Add $\square$ to each side.

$r + 0 = \square$

$r = \square$

Check:

$r + (-13) = 15$

$\square + (-13) = 15$ — Replace with $\square$.

$\square = 15$ — $28 + (-13) = \square$

2 $-4.8 + y = -13.7$

$-4.8 + y = -13.7$

$-4.8 + y + \square = -13.7 + \square$ — Add $\square$ to each side.

$y + 0 = \square$

$y = \square$ — Check your solution.

Your Turn **Solve each equation. Check your solution.**

a. $z - 12 = 5$

b. $8 + m = -6$

EXAMPLES

3 Solve $k + 12 = -6$.

$$k + 12 = -6$$

$k + 12 - \square = -6 - \square$ Subtract $\square$ from each side.

$k = \square$ Check your solution.

4 Solve $x - \left(-\frac{2}{5}\right) = \frac{7}{10}$.

$x + \square = \frac{7}{10}$ Rewrite the equation.

$x + \frac{2}{5} - \square = \frac{7}{10} - \square$ Subtract $\square$ from each side.

$x = \frac{7}{10} - \square$ Rewrite $\frac{2}{5}$ as $\square$.

$x = \square$ Check your solution.

Your Turn **Solve each equation. Check your solution.**

a. $y + 10 = -2$

b. $x - \left(-\frac{3}{4}\right) = \frac{11}{12}$

HOMEWORK ASSIGNMENT

Page(s):

Exercises:

3–7 Solving Equations Involving Absolute Value

WHAT YOU'LL LEARN

- Solve equations involving absolute value.

FOLDABLES™

ORGANIZE IT

Write an equation that contains an absolute value under the tab for Absolute Value. Solve the equation using methods 1 and 2.

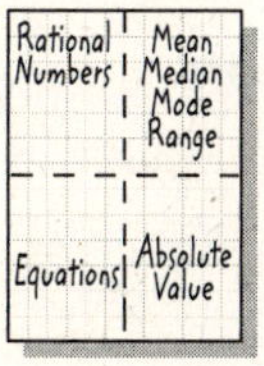

EXAMPLES

1 **Solve $|d - 4| = 3$. Check your solution.**

Method 1: Use a number line

$|d - 4| = 3$ means the distance between d and 4 is ____ units. So, to find x on the number line, start at 4 and move ____ units in either direction.

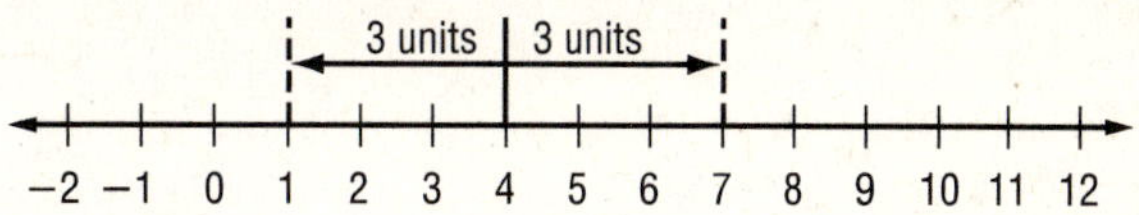

$d =$ ____ $\qquad d =$ ____

Method 2: Write and solve a compound sentence.

$|d - 4| = 3$ also means $d - 4 =$ ____ or $d - 4 =$ ____.

$d - 4 =$ ____	or	$d - 4 =$ ____
$d - 4 +$ ____ $= 3 +$ ____		$d - 4 +$ ____ $= -3 +$ ____
$d =$ ____		$d =$ ____

Check:

Replace d with ____.	Replace d with ____.
$\|d - 4\| = 3$	$\|d - 4\| = 3$
$\|$ ____ $- 4\| = 3$	$\|$ ____ $- 4\| = 3$
$\|$ ____ $\| = 3$	$\|$ ____ $\| = 3$
____ $= 3$	____ $= 3$

The solution set is ____.

2 **Solve $|g + 3| - 2 = 6$**

To solve the equation, first rewrite the equation.

$|g + 3| - 2 = 6$

$|g + 3| - 2 + \square = 6 + \square$ Add $\square$ to each side.

$|g + 3| = \square$

Next, write a compound sentence and solve it.

$g + 3 = \square$ or $g + 3 = \square$

$g + 3 - \square = 8 - \square$ $g + 3 - \square = -8 - \square$

$g = \square$ $g = \square$

The solution set is $\square$.

Your Turn **Solve and check your solution.**

a. $|a - 2| = 5$

b. $|x + 4| - 3 = 8$

WRITE IT

Write an example of an absolute value equation that has no solution.

BUILD YOUR VOCABULARY (page 48)

A set with $\square$ members is called an **empty set**.

EXAMPLE

3 **Solve $|y + 5| - 2 = -7$.**

First simplify the expression.

$|y + 5| - 2 = -7$

$|y + 5| - 2 \square = -7 \square$ Add $\square$ to each side.

$|y + 5| = -5$ $|y + 5| = -5$ is never true.

The solution is the $\square$ or $\square$.

Your Turn Solve $|x + 3| - 1 = -8$.

HOMEWORK ASSIGNMENT

Page(s):

Exercises:

CHAPTER 3

BRINGING IT ALL TOGETHER

STUDY GUIDE

FOLDABLES™	VOCABULARY PUZZLEMAKER	BUILD YOUR VOCABULARY
Use your **Chapter 3 Foldable** to help you study for your chapter test.	To make a crossword puzzle, word search, or jumble puzzle of the vocabulary words in Chapter 3, go to: www.glencoe.com/sec/math/t_resources/free/index.php	You can use your completed **Vocabulary Builder** (pages 48–49) to help you solve the puzzle.

3-1 Rational Numbers

1. Explain why $\frac{-3}{7}$, $0.\overline{6}$ and 15 are rational numbers.

Match the correct inequality symbols at the right with their correct description on the left.

2. less than or equal to

3. greater than

4. less than

a. $>$
b. $\geq$
c. $<$
d. $\leq$

3-2 Adding and Subtracting Rational Numbers

Find each sum or difference.

5. $-2.6 + 5.8$

6. $7.3 - 12.4$

7. $-10.1 - (-14.9)$

8. $\frac{1}{4} - \left(-\frac{1}{8}\right)$

9. $\frac{5}{6} + \left(-\frac{7}{8}\right)$

10. $-5\frac{1}{8} + \left(-2\frac{15}{16}\right)$

3-3 Mean, Median, Mode, and Range

Match the measure of central of tendency at the right with its description on the left

11. the score used most often

12. the average score

13. the middle score

a. mean
b. median
c. mode
d range

Use the stem-and-leaf plot shown.

14. What is the median of the data set?

15. What is the mean? Round to the nearest tenth if necessary.

Stem	Leaf
2	0 1 1 2 5
3	2 2 2 7 9 9
4	1 3 3
5	6 6 8 9
6	0 1 8 8 8 $4\|2 = 42$

16. What is the mode?

3-4 Equations

17. How can you tell whether an equation is an open sentence?

18. Consider the equation $3n + 6 = 15$ with the replacement set {0, 1, 2, 3, 4, 5}. Complete the chart to find the solution(s) of the equation.

Value for n	$3n + 6 = 15$	True or False?
0		
1		
2		
3		
4		
5		

3-5 Solving Equations by Using Models

19. For each algebra tile, write the part of the equation that it represents.

a. [x]

b. [−]

Solve each equation.

20. $-3 + m = 7$

21. $z + 2 = -5$

22. $n - 4 = -1$

23. $6 = k + 1$

3-6 Solving Addition and Subtraction Equations

24. To solve $x + 17 = 46$ using the Subtraction Property of Equality, you would subtract ______ from each side.

25. To solve $y - 9 = -30$ using the Addition Property of Equality, you would add ______ to each side.

26. Write an equation that you could solve by subtracting 32 from each side.

3-7 Solving Equations Involving Absolute Value

Determine whether each sentence is *sometimes*, *always*, or *never* true. Explain.

27. $|x - 3| = 12$

28. $-3 = |m + (-6)|$

29. $|n| - 8 = 4$

30. $-6 = |y| + 7$

CHAPTER 3

Checklist

ARE YOU READY FOR THE CHAPTER TEST?

Math Online

Visit **algconcepts.com** to access your textbook, more examples, self-check quizzes, and practice tests to help you study the concepts in Chapter 3.

Check the one that applies. Suggestions to help you study are given with each item.

☐ **I completed the review of all or most lessons without using my notes or asking for help.**

- You are probably ready for the Chapter Test.
- You may want take the Chapter 3 Practice Test on page 135 of your textbook as a final check.

☐ **I used my Foldable or Study Notebook to complete the review of all or most lessons.**

- You should complete the Chapter 3 Study Guide and Review on pages 132–134 of your textbook.
- If you are unsure of any concepts or skills, refer back to the specific lesson(s).
- You may also want to take the Chapter 3 Practice Test on page 135.

☐ **I asked for help from someone else to complete the review of all or most lessons.**

- You should review the examples and concepts in your Study Notebook and Chapter 3 Foldable.
- Then complete the Chapter 3 Study Guide and Review on pages 132–134 of your textbook.
- If you are unsure of any concepts or skills, refer back to the specific lesson(s).
- You may also want to take the Chapter 3 Practice Test on page 135.

Student Signature

Parent/Guardian Signature

Teacher Signature

Chapter 4

Multiplication and Division Equations

FOLDABLES™ Use the instructions below to make a Foldable to help you organize your notes as you study the chapter. You will see Foldable reminders in the margin of this Interactive Study Notebook to help you in taking notes.

Begin with seven sheets of grid paper.

STEP 1 **Fold**
Fold each sheet in half along the width.

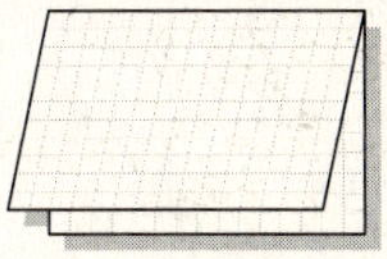

STEP 2 **Unfold**
Unfold and cut four rows from the left side of each sheet, from the top to the crease.

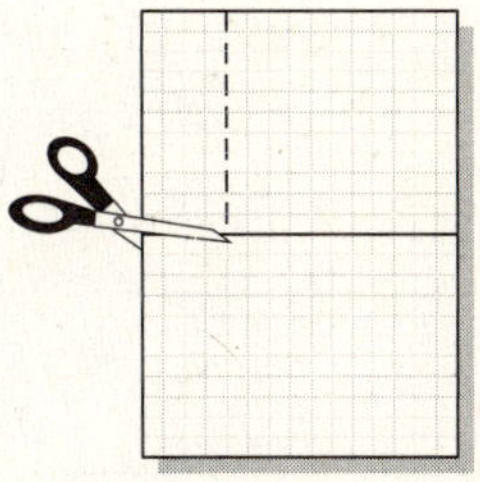

STEP 3 **Stack**
Stack the sheets and staple to form a booklet.

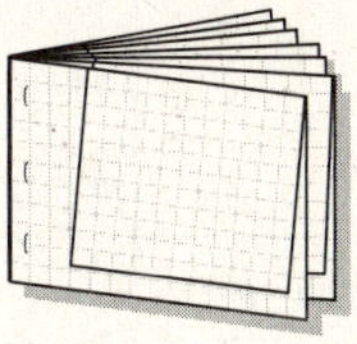

STEP 4 **Label**
Label each page with a lesson number and title.

NOTE-TAKING TIP: When you take notes, record important ideas and examples from each lesson. Be sure to include examples that will help you understand the concepts.

Chapter 4

CHAPTER 4

BUILD YOUR VOCABULARY

This is an alphabetical list of new vocabulary terms you will learn in Chapter 4. As you complete the study notes for the chapter, you will see Build Your Vocabulary reminders to complete each term's definition or description on these pages. Remember to add the textbook page number in the second column for reference when you study.

Vocabulary Term	Found on Page	Definition	Description or Example
combination			
consecutive integers [con-SEC-yoo-tiv]			
event			
Fundamental Counting Principle			
factorial [fak-TOR-ee-ul]			
grouping symbols			
identity			
multiplicative inverses [mul-tah-PLIK-uh-tiv]			
outcomes			
permutation [PUR-myu-TAY-shun]			
reciprocal [ree-SIP-ruh-kul]			
sample space			
tree diagram			

4–1 Multiplying Rational Numbers

WHAT YOU'LL LEARN

- Multiply rational numbers.

KEY CONCEPTS

Multiplying Two Rational Numbers with Different Signs The product of two rational numbers with different signs is negative.

Multiplying Two Rational Numbers with the Same Sign The product of two rational numbers with the same sign is positive.

EXAMPLES

Find each product.

1 $-3.2(5)$

$-3.2(5) =$ ______ The factors have different signs.

The product is ______.

2 $-4.7(-0.4)$

$-4.7(-0.4) =$ ______ The factors have the same sign.

The product is ______.

Your Turn **Find each product.**

a. $-8.5(3)$

b. $-9.1(-0.8)$

EXAMPLE

3 **A skydiver jumps from 12,000 feet. Solve the equation $h = 12{,}000 + (0.5)(-32.1)(144)$ to find the skydiver's height after he free-falls for 12 seconds.**

$h = 12{,}000 + (0.5)(-32.1)(144)$

$= 12{,}000 +$ ______ (144) Multiply 0.5 and −32.1.

$= 12{,}000 +$ ______ Multiply −16.05 and 144.

$=$ ______ Add 12,000 and −2311.2.

After 12 seconds, the skydiver's height is ______ feet.

Your Turn A skydiver jumps 12,000 feet. Solve the equation $h = 12{,}000 + (0.5)(-32.1)(196)$ to find the skydiver's height after he free-falls for 14 seconds.

EXAMPLES

KEY CONCEPT

Multiplying Fractions
To multiply fractions, multiply the numerators and multiply the denominators.

FOLDABLES™ Write the rules for multiplying rational numbers in your notes. Be sure to include examples.

Find each product.

4 $\frac{2}{7}\left(-\frac{3}{5}\right)$

$\frac{2}{7}\left(-\frac{3}{5}\right) = -\frac{\square}{\square}$ Multiply the numerators and multiply the denominators.

$= \square$ The factors have different signs.

The product is ______.

5 $-8 \cdot \left(-\frac{3}{7}\right)$

$-8 \cdot \left(-\frac{3}{7}\right) = \square \cdot \left(-\frac{3}{7}\right)$ Rewrite −8 as an improper fraction.

$= \left(\frac{8 \cdot 3}{1 \cdot 7}\right)$ Multiply the numerators and multiply the denominators.

$= \square$ or $\square$ The factors have the same sign. The product is ______.

6 $-3\frac{1}{3} \cdot \frac{1}{7}$

$-3\frac{1}{3} \cdot \left(\frac{1}{7}\right) = \square \cdot \frac{1}{7}$ Rewrite $-3\frac{1}{3}$ as an improper fraction.

$= \frac{-10 \cdot 1}{3 \cdot 7}$ Multiply the numerators and multiply the denominators.

$= \square$ The factors have different signs.

The product is ______.

Your Turn **Find each product.**

a. $-\frac{2}{3} \cdot \frac{4}{9}$

b. $-\frac{3}{4} \cdot (-7)$

c. $-2\frac{1}{2} \cdot \left(-\frac{5}{6}\right)$

EXAMPLE

KEY CONCEPT

Multiplicative Property of –1 The product of –1 and any number is the number's additive inverse.

7 **Simplify $5b(-2.2y)$.**

$5b(-2.2y) = (5)(b)(-2.2)(y)$	$5b = (5)(b)$; $(-2.2y) = (-2.2)(y)$
$= (5)(-2.2)$ ______	Commutative Property
$=$ ______ $(b \cdot y)$	Associative Property
$=$ ______	Simplify.

Your Turn Simplify $-\frac{1}{4}p\left(-\frac{3}{5}r\right)$.

HOMEWORK ASSIGNMENT

Page(s):

Exercises:

4–2 Counting Outcomes

BUILD YOUR VOCABULARY (page 70)

Outcomes are all possible __________ of a counting problem or the results of an experiment.

A **tree diagram** is used to show the total number of possible __________.

The **sample space** is the list of all possible outcomes.

WHAT YOU'LL LEARN

- Use tree diagrams and the Fundamental Counting Principle to count outcomes.

FOLDABLES ORGANIZE IT

Write an example of a counting problem and draw a tree diagram for it under the tab for Counting Outcomes.

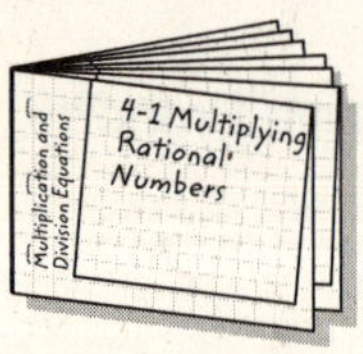

EXAMPLE

1 **Brooke is shopping for a new computer system. She has a list of 2 different CPUs, 3 different monitors, and 3 different printers. How many different ways can she choose one CPU, one monitor, and one printer?**

Make a tree diagram to find the number of combinations.

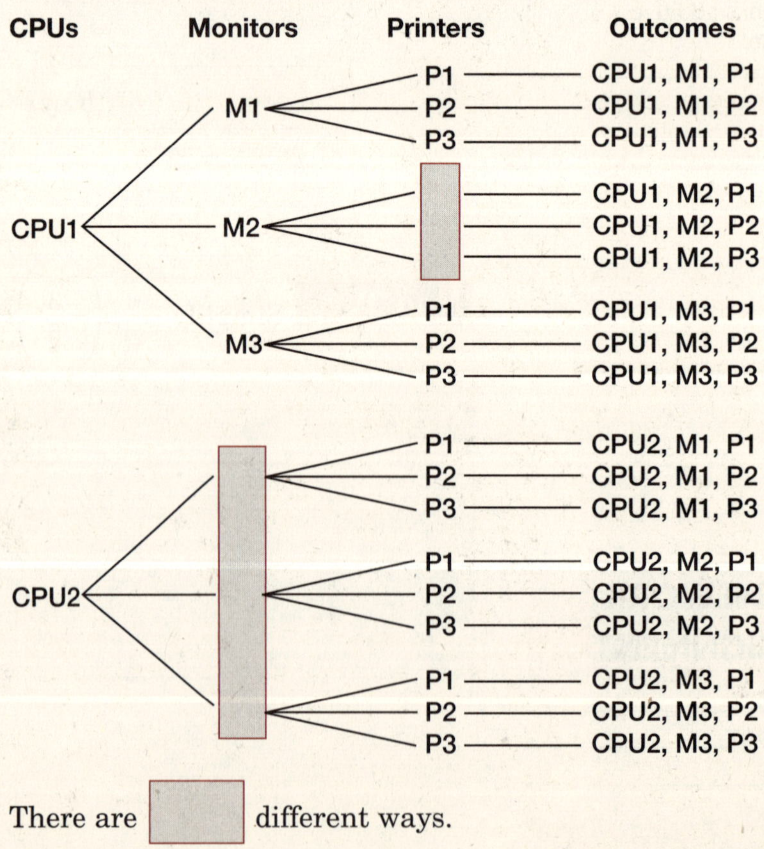

There are ______ different ways.

Your Turn Suppose you can order a pizza with 2 different types of crust, 2 different types of sauce, and 4 different types of toppings. How many ways can you order a pizza?

BUILD YOUR VOCABULARY (page 70)

An **event** is the subset of the possible outcomes, or ______.

EXAMPLE

KEY CONCEPT

Fundamental Counting Principle If event *M* can occur in *m* ways and is followed by event *N* that can occur in *n* ways, then the event *M* followed by event *N* can occur in $m \times n$ ways.

2 **How many different kinds of photo processing are possible?**

Process Time	Paper Type	Photo Size
1 hour 1 day	regular glossy deluxe	3 by 5 4 by 6

There are ____ processing times, ____ paper types, and ____ photo sizes, so the number of different kinds of photo processing is ____ × ____ × ____ or ____.

Your Turn A concession stand offers the choices shown in the table below. Suppose one item is selected from each column. How many different choices are possible?

Meat	Topping	Bun
hot dog hamburger veggie burger	ketchup mustard onions	wheat white

HOMEWORK ASSIGNMENT

Page(s):

Exercises:

4–3 Dividing Rational Numbers

WHAT YOU'LL LEARN

- Divide rational numbers.

KEY CONCEPTS

Dividing Rational Numbers with Different Signs or the Same Sign The quotient of two numbers with different signs is negative.

The quotient of two numbers with the same sign is positive.

Multiplicative Inverse Property The product of a number and its multiplicative inverse is 1.

Dividing Fractions To divide a fraction by any nonzero number, multiply by the reciprocal of a number.

EXAMPLES

Find each quotient.

1 $\mathbf{8 \div (-2.5)}$

$8 \div (-2.5) =$ ____ Numbers have different signs.

The quotient is ____.

2 $\mathbf{-9.3 \div (-0.3)}$

$-9.3 \div (-0.3) =$ ____ Numbers have the ____ sign. The quotient is positive.

Your Turn **Find each quotient.**

a. $-14.4 \div (-0.6)$

b. $8.8 \div (-1.1)$

BUILD YOUR VOCABULARY (page 70)

Two numbers whose product is ____ are called **multiplicative inverses** or **reciprocals.**

EXAMPLES

Find each quotient.

3 $\mathbf{-12 \div \frac{2}{5}}$

$-12 \div \frac{2}{5} = -12 \cdot$ ____ To divide by $\frac{2}{5}$, multiply by its reciprocal.

$= -\frac{60}{2}$ The numbers have different signs. The product is negative.

$=$ ____

4 $2\frac{3}{7} \div \frac{1}{2}$

$2\frac{3}{7} \div \frac{1}{2} = \_\_\_\_ \div \frac{1}{2}$ — Rewrite $2\frac{3}{7}$ as an improper fraction.

$= \_\_\_\_ \cdot \_\_\_\_$ — To divide by $\frac{1}{2}$, multiply by its reciprocal.

$= \_\_\_\_$ or $\_\_\_\_$ — The numbers have the same sign.

The product is ______.

REMEMBER IT

If one of the numbers in a division problem is an integer, write it as a fraction with a denominator of 1.

Your Turn **Find each quotient.**

a. $-\frac{3}{8} \div (-6)$

b. $3\frac{1}{2} \div \left(-\frac{3}{4}\right)$

EXAMPLE

FOLDABLES™

ORGANIZE IT

Under the tab for Dividing Rational Numbers, write an example showing how to divide two fractions.

5 **Two paintings are to be hung on a wall so that the distances between the centers of the paintings is the same as the distance from either center to the end of the wall. The paintings are both 6 feet wide and the wall measures $22\frac{1}{4}$ feet across. How far from the end of the wall closest to it should the center of each painting be located?**

Draw a picture of the 2 paintings and the walls on either side. Draw a dotted line down the center of each painting. Notice how this divides the wall into 3 pieces. Divide the wall length by 3.

$22\frac{1}{4} \div 3 = \_\_\_\_ \div 3$

$= \_\_\_\_ \cdot \_\_\_\_$

$= \_\_\_\_$ or $\_\_\_\_$

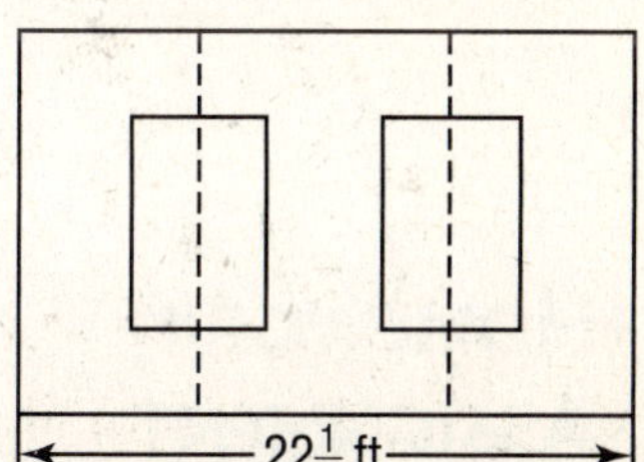

The center of each painting should be ______ feet from the wall.

Your Turn Larry needs to saw a board into 3 equal pieces. The board measures $18\frac{3}{4}$ inches long. How many inches from each end should he make the cuts?

EXAMPLE

6 Evaluate $\frac{3}{x}$ if $x = -\frac{3}{4}$.

$\frac{3}{x} = \frac{3}{\square}$	Replace x with $\square$.
$= 3 \div \square$	Rewrite the fraction as a division sentence.
$= \frac{3}{1} \cdot \square$	Replace 3 by $\frac{3}{1}$. To divide by $-\frac{3}{4}$, multiply by its reciprocal.
$= -\frac{12}{3}$	Multiply the numerators and multiply the denominators.
$= \square$	Simplify.

Your Turn Evaluate $\frac{x}{5}$ if $x = -\frac{2}{7}$.

HOMEWORK ASSIGNMENT

Page(s):

Exercises:

4–4 Solving Multiplication and Division Equations

WHAT YOU'LL LEARN

- Solve multiplication and division equations by using the properties of equality.

KEY CONCEPT

Division Property of Equality If you divide each side of an equation by the same nonzero number, the two sides remain equal.

EXAMPLES

Solve each equation. Check your solution.

1 $\mathbf{5b = 30}$

$5b = 30$

$\frac{5b}{\square} = \frac{30}{\square}$ Divide each side by ☐.

$b = \square$

Check: $5b = 30$

$5\square = 30$ Substitute $b =$ ☐.

$\square = 30$

2 $\mathbf{-24 = 3g}$

$-24 = 3g$

$\frac{-24}{\square} = \frac{3g}{\square}$ Divide each side by ☐.

$\square = g$ Check the solution.

3 $\mathbf{-5.5z = -22}$

$-5.5z = -22$

$\frac{-5.5z}{\square} = \frac{-22}{\square}$ Divide each side by ☐.

$z = \square$ Check the solution.

Your Turn **Solve each equation. Check your solution.**

a. $-3x = 39$

WRITE IT

Explain how you can recognize when to divide both sides of an equation by a number.

______

______

______

______

b. $-18t = -90$

c. $4.2a = -21$

EXAMPLE

4 **Brian received a $25 gift certificate from his grandparents for his birthday. How many $2.35 packages of trading cards can he buy with the gift certificate?**

Let c represent the number of packages of trading cards. Write an equation to represent the problem.

Price per package	times	number of packages	equals	total cost.
☐	$\times$	c	$=$	☐

Solve the equation for c.

$$\square c = \square$$

$$\frac{2.35c}{\square} = \frac{25}{\square}$$

$$c \approx \square$$

Since Brian cannot buy part of a package, he has enough money to buy ☐ packages.

Your Turn Allison is planning her birthday party. She wants to take her guests to a movie that costs $7.25 per person. She has $40 to spend on her guests. How many guests can she invite?

REVIEW IT

Any word problem can be solved using the four-step plan. List the steps below. *(Lesson 1-5)*

______

______

______

______

KEY CONCEPT

Multiplication Property of Equality If you multiply each side of an equation by the same number, the two sides remain equal.

FOLDABLES™

ORGANIZE IT

Write examples showing how to solve multiplication or division equations under the tab for Solving Multiplication and Division Equations.

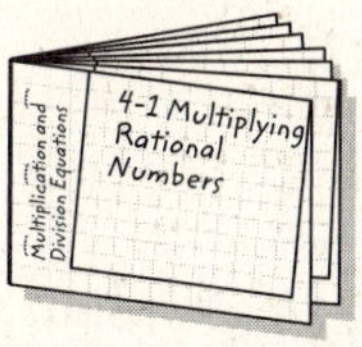

HOMEWORK ASSIGNMENT

Page(s):

Exercises:

EXAMPLES

Solve each equation.

5 $\frac{w}{7} = 6$

$\frac{w}{7} = 6$

$\square\left(\frac{w}{7}\right) = \square(6)$ — Multiply each side by $\square$.

$w = \square$ — Check the solution.

6 $-9 = \frac{1}{2}m$

$-9 = \frac{1}{2}m$

$\square(-9) = \square\left(\frac{1}{2}m\right)$ — Multiply each side by $\square$.

$\square = m$ — Check the solution.

7 $-\frac{2}{5}x = -8$

$-\frac{2}{5}x = -8$

$\square\left(-\frac{2}{5}x\right) = \square(-8)$ — Multiply each side by $\square$.

$x = \square$ — Check the solution.

Your Turn **Solve each equation. Check your solution.**

a. $\frac{m}{-5} = 12$

b. $-\frac{1}{3}b = 7$

c. $24 = -\frac{3}{4}y$

4–5 Solving Multi-Step Equations

EXAMPLES

WHAT YOU'LL LEARN

- Solve equations involving more than one operation.

FOLDABLES™

ORGANIZE IT

Write and solve an equation that involves multiple steps. Write your equation under the tab for Solving Multi-Step Equations.

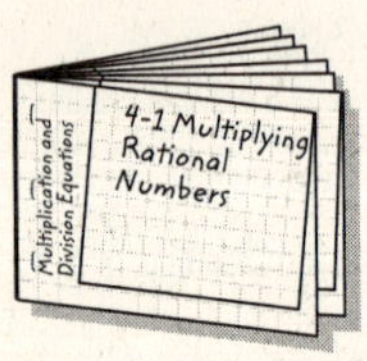

Solve each equation.

1 $\frac{x}{6} - 4 = 2$

$$\frac{x}{6} - 4 = 2$$

$$\frac{x}{6} - 4 + \_\_\_\_ = 2 + \_\_\_\_$$ Add ____ to each side.

$$\frac{x}{6} = 6$$

$$\_\_\_\_\left(\frac{x}{6}\right) = \_\_\_\_(6)$$ Multiply each side by ____.

$$x = \_\_\_\_$$ Check the solution.

2 $3m + 12 = 27$

$$3m + 12 = 27$$

$$3m + 12 - \_\_\_\_ = 27 - \_\_\_\_$$ Subtract ____ from each side.

$$3m = 15$$

$$\frac{3m}{\_\_\_\_} = \frac{15}{\_\_\_\_}$$ Divide each side by ____.

$$m = \_\_\_\_$$ Check the solution.

Your Turn **Solve each equation. Check your solution.**

a. $\frac{h}{5} + 6 = 13$

b. $9 - 4a = 45$

EXAMPLE

3 Solve $-6.2 = \frac{3 + n}{15}$.

$$-6.2 = \frac{3+n}{15}$$

$$15(-6.2) = 15\left(\frac{3+n}{15}\right)$$ Multiply each side by 15.

$$-93 = \underline{\hspace{2cm}}$$

$$-93 - \underline{\hspace{1cm}} = 3 - \underline{\hspace{1cm}} + n$$ Subtract.

$$\underline{\hspace{1.5cm}} = n$$

Your Turn Solve $\frac{-x + 3}{7} = -3$. Check your solution.

EXAMPLE

Remember It

When solving word problems, always check to make sure your answer is reasonable. Ask yourself: Does this answer make sense?

4 In a city, the tallest building is 1268 feet tall. This is 35 feet greater than 3 times the height of the fifth tallest building. How tall is the fifth tallest building?

Let x represent the height of the fifth tallest building. Translate the information into an equation and solve.

Height of tallest building	equals	35 feet	plus	3 times the height of the fifth tallest building.
1268	=	35	+	$3x$

$$1268 = 35 + 3x$$

$$1268 - \underline{\hspace{1cm}} = 35 + 3x - \underline{\hspace{1cm}}$$ Subtract ______ from each side.

$$1233 = 3x$$

$$\frac{1233}{\underline{\hspace{1cm}}} = \frac{3x}{\underline{\hspace{1cm}}}$$ Divide each side by ______.

$$\underline{\hspace{1.5cm}} = x$$

The fifth tallest building is ______ feet tall.

Your Turn The Parker family recently purchased a new car. Their old car had 105,000 miles on its odometer. This is 50,000 more than four times the number of miles on the new car. How many miles does the new car have on its odometer?

BUILD YOUR VOCABULARY (page 70)

Consecutive integers are integers in counting order such as 3, 4, and 5.

EXAMPLE

5 **Find four consecutive odd integers whose sum is −8.**

Let n represent the first odd integer. Then $n + 2$ represents the second odd integer, $n + 4$ represents the third, and $n + 6$ represents the fourth.

$$\square + \square + \square + \square = -8$$

$$4n + 12 = -8$$

$$4n + 12 - \square = -8 - \square$$

$$4n = -20$$

$$\frac{4n}{\square} = \frac{-20}{\square}$$

$$n = \square$$

The numbers are $\square$, $n + 2$ or $\square$, $n + 4$ or $\square$, and $n + 6$ or $\square$.

HOMEWORK ASSIGNMENT

Page(s):

Exercises:

Your Turn Find four consecutive even integers whose sum is −28.

4–6 Variables on Both Sides

WHAT YOU'LL LEARN

- Solve equations with variables on both sides.

FOLDABLES™

ORGANIZE IT

Write and solve an equation that involves variables on both sides of the equation. Write your equation under the tab for Variables on Both Sides.

EXAMPLE

Solve each equation.

1 $\mathbf{y + 8 = 9y}$

$$y + 8 = 9y$$

$$y + 8 - \square = 9y - \square$$ Subtract $\square$ from each side.

$$8 = 8y$$

$$\frac{8}{\square} = \frac{8y}{\square}$$ Divide each side by $\square$.

$$\square = y$$

Your Turn **Solve each equation.**

a. $4x = 10x - 3$

b. $\frac{1}{3}t + 4 = -\frac{2}{3}t$

BUILD YOUR VOCABULARY (page 70)

An ______ that is ______ for every value of the variable is called an **identity**.

EXAMPLES

Solve each equation.

2 $\mathbf{3 - 4y = 6 - 4y}$

$$3 - 4y = 6 - 4y$$

$$3 - \square + 4y = 6 - 4y + \square$$ Add $\square$ to each side.

$$3 = 6$$

The equation has ______. $3 = 6$ is never true.

3 $\mathbf{8m + 2 = 2 - 3m + 11m}$

$8m + 2 = 2 - 3m + 11m$

$8m + 2 = 2 +$ ______ $\quad -3m + 11m =$ ______

$8m + 2 = 8m + 2$ ______ Property

The equation is an ______. $8m + 2 = 8m + 2$ is true for all values of m.

Your Turn **Solve each equation.**

a. $9 - 5h + 4 = -5h + 13$

b. $3y + 11 - 8y = 4 - 5y + 6$

HOMEWORK ASSIGNMENT

Page(s):

Exercises:

4–7 Grouping Symbols

What You'll Learn

- Solve equations with grouping symbols.

Build Your Vocabulary (page 70)

Symbols that group terms together in an expression or equation are called **grouping symbols**. Parentheses and brackets are examples of grouping symbols.

Foldables™

Organize It

Write and solve an equation that involves grouping symbols. Write your equation under the tab for Grouping Symbols.

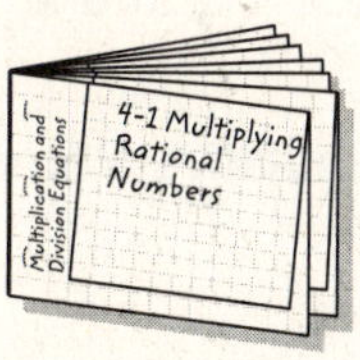

EXAMPLES

Solve each equation. Check your solution.

1 $\mathbf{5(2x - 1) = -25}$

$5(2x - 1) = -25$

$10x - 5 = -25$ Distributive Property

$10x - 5 + \square = -25 + \square$ Add $\square$ to each side.

$10x = -20$

$\dfrac{10x}{\square} = \dfrac{-20}{\square}$ Divide.

$x = -2$

2 $\mathbf{5(h + 6) - 6 = 3(5h - 2)}$

$5(h + 6) - 6 = 3(5h - 2)$

$\square - 6 = \square$ Distributive Property

$5h + \square = 15h - 6$ Add like terms.

$5h + 24 - \square = 15h - 6 - \square$ Subtract.

$24 = 10h - 6$

$24 + \square = 10h - 6 + 6$ Add $\square$ to each side.

$30 = 10h$

$\dfrac{30}{\square} = \dfrac{10h}{\square}$ Divide.

$3 = h$ Check the solution.

Remember It

When you add integers with the same sign, add their absolute values. The sign is the same as the sign of the integers.

When you add integers with different signs, first find the difference of their absolute values. The sign will be the same as the sign of the integer with the greater absolute value. *(Lesson 2-3)*

Your Turn **Solve each equation. Check your solution.**

a. $4(3x + 7) = -8$

b. $7(3 - p) - 7 = -4(p - 5)$

EXAMPLE

3 The area of the trapezoid is 64 square millimeters. Find the value of x.

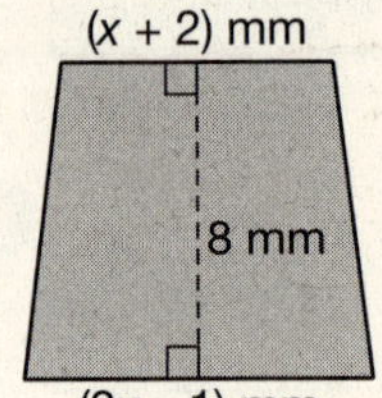

$A = \frac{1}{2}(b_1 + b_2)h$ Area of a trapezoid

$\square = \frac{1}{2}[\square + \square]\square$

$64 = \frac{1}{2}(3x + 1)8$ Add $2x - 1$ and $x + 2$.

$64 = \frac{1}{2} \cdot 8 \cdot (3x + 1)$ Commutative Property

$64 = \square(3x + 1)$

$64 = 12x + 4$ Distributive Property

$64 - \square = 12x + 4 - \square$ Subtract.

$60 = 12x$

$\frac{60}{\square} = \frac{12x}{\square}$ Divide.

$\square = x$

The value of x is $\square$.

Your Turn The area of a trapezoid is 52 square inches. Find the lengths of the bases if one base is 2 inches more than the other base, and the altitude is 4 inches.

Homework Assignment

Page(s):

Exercises:

CHAPTER 4

BRINGING IT ALL TOGETHER

STUDY GUIDE

FOLDABLES	VOCABULARY PUZZLEMAKER	BUILD YOUR VOCABULARY
Use your **Chapter 4 Foldable** to help you study for your chapter test.	To make a crossword puzzle, word search, or jumble puzzle of the vocabulary words in Chapter 4, go to: www.glencoe.com/sec/math/t_resources/free/index.php	You can use your completed **Vocabulary Builder** (page 70) to help you solve the puzzle.

4-1 Multiplying Rational Numbers

Multiply.

1. $-6.2 \cdot (-0.3)$

2. $4(-8.2)$

3. $\frac{-4}{3}(6)$

4. $\frac{-3}{8} \cdot \frac{-5}{7}$

4-2 Counting Outcomes

Kynda is playing in a chess tournament. The tree diagram for her possible outcomes is shown below.

5. Name two different outcomes.

6. Three different outcomes result in a win-loss record of 2-1. What are they?

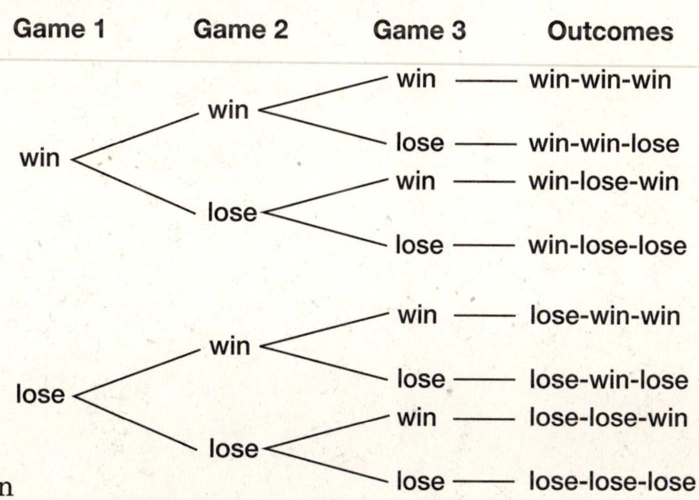

7. How many outcomes are possible in choosing a sundae with 7 different types of ice cream, 4 different types of toppings, and 3 different types of sauce?

4-3

Dividing Rational Numbers

Write *negative* or *positive* to describe each quotient. Then find the quotient.

8. $\frac{48.6}{16.2}$

9. $-13.4 \div 2$

10. $\frac{3}{7} \div \left(-\frac{4}{5}\right)$

11. $\frac{-5}{6} \div \left(-\frac{8}{9}\right)$

4-4

Solving Multiplication and Division Equations

Complete the sentence after each equation to tell how you would solve the equation.

12. $\frac{x}{7} = 16$

____ each side by ____.

13. $5x = 125$

____ each side by ____, or multiply each side by ____.

14. $-8k = 96$

Divide each side by ____, or multiply each side by ____.

15. Explain how rewriting $4\frac{1}{3}x = 2\frac{1}{8}$ as $\frac{13}{3}x = \frac{17}{8}$ helps you solve the equation.

4-5

Solving Multi-Step Equations

16. Suppose you want to solve $\frac{x+3}{5} = 6$.

a. What is the first step in solving the equation?

b. What is the next step in solving the equation?

17. The sum of two consecutive odd integers is −36.

a. Write an equation for this situation.

b. What are the two consecutive odd integers?

4-6

Variables on Both Sides

State the first step in solving each equation.

18. $-3x + 6 = -10x + 10$

19. $-3.6 + 4.2z = 3 - 2.1z$

20. $\frac{1}{9}y - 3 = \frac{5}{9}y$

21. Suppose you want to help a friend solve $6k + 7 = 3k - 8$. What would you advise her to do first? Why?

4-7

Grouping Symbols

Solve each equation. Check your solution.

22. $-8(x - 5) = 4(2 - x)$

23. $2(x + 1) - 3(x + 2) = 7$

CHAPTER 4

Checklist

ARE YOU READY FOR THE CHAPTER TEST?

Math Online

Visit **algconcepts.com** to access your textbook, more examples, self-check quizzes, and practice tests to help you study the concepts in Chapter 4.

Check the one that applies. Suggestions to help you study are given with each item.

☐ **I completed the review of all or most lessons without using my notes or asking for help.**

- You are probably ready for the Chapter Test.
- You may want take the Chapter 4 Practice Test on page 183 of your textbook as a final check.

☐ **I used my Foldable or Study Notebook to complete the review of all or most lessons.**

- You should complete the Chapter 4 Study Guide and Review on pages 180–182 of your textbook.
- If you are unsure of any concepts or skills, refer back to the specific lesson(s).
- You may also want to take the Chapter 4 Practice Test on page 183.

☐ **I asked for help from someone else to complete the review of all or most lessons.**

- You should review the examples and concepts in your Study Notebook and Chapter 4 Foldable.
- Then complete the Chapter 4 Study Guide and Review on pages 180–182 of your textbook.
- If you are unsure of any concepts or skills, refer back to the specific lesson(s).
- You may also want to take the Chapter 4 Practice Test on page 183.

Student Signature

Parent/Guardian Signature

Teacher Signature

CHAPTER 5

Proportional Reasoning and Probability

FOLDABLES™ Use the instructions below to make a Foldable to help you organize your notes as you study the chapter. You will see Foldable reminders in the margin of this Interactive Study Notebook to help you in taking notes.

Begin with a sheet of notebook paper.

STEP 1 **Fold**
Fold lengthwise to the holes.

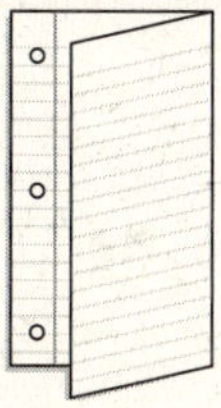

STEP 2 **Cut**
Cut four tabs.

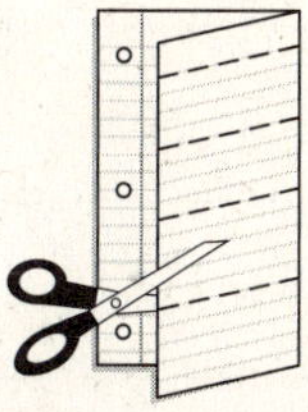

STEP 3 **Label**
Label the tabs using lesson concepts as shown.

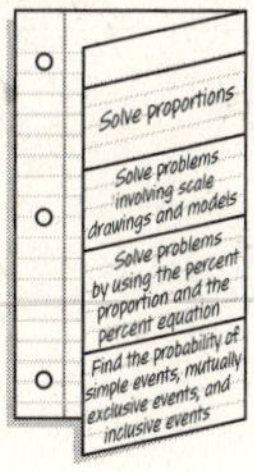

NOTE-TAKING TIP: When you take notes, it helps to write about when you would use the concept in your daily life. For example, percents are used when calculating the tax on an item purchased.

CHAPTER 5

BUILD YOUR VOCABULARY

This is an alphabetical list of new vocabulary terms you will learn in Chapter 5. As you complete the study notes for the chapter, you will see Build Your Vocabulary reminders to complete each term's definition or description on these pages. Remember to add the textbook page number in the second column for reference when you study.

Vocabulary Term	Found on Page	Definition	Description or Example
base			
box-and-whisker plot			
circle graph			
complement [kahm-PLU-ment]			
compound event			
dimensional analysis [duh-MEN-shun-ul]			
empirical probability [im-PEER-i-kul]			
experimental probability [ek-speer-uh-MEN-tul]			
extremes			
inclusive [in-KLOO-siv]			
independent events			
lower quartile [KWAR-tile]			
mutually exclusive [MYOO-chew-a-lee]			
odds			

Vocabulary Term	Found on Page	Definition	Description or Example
percent equation			
percent of decrease			
percent of increase			
percent proportion			
percentage			
percentile			
probability [PRA-buh-BIL-i-tee]			
proportion [pro-POR-shun]			
random			
rate			
ratio			
scale			
scale drawing			
scale model			
theoretical probability [thee-uh-RET-i-kul]			
unit rate			
upper quartile			

5–1 Solving Proportions

BUILD YOUR VOCABULARY (page 95)

A **ratio** is the ____ of two numbers.

An equation stating that two ratios are ____ is a **proportion**.

WHAT YOU'LL LEARN

- Solve proportions.

KEY CONCEPT

Property of Proportions The cross products of a proportion are equal.

EXAMPLES

Solve each proportion.

1 $\dfrac{9}{m} = \dfrac{63}{35}$

$$\frac{9}{m} = \frac{63}{35}$$

$9\,$____$\, = 63\,$____ Find the cross products.

$315 = 63m$

$\dfrac{315}{\text{____}} = \dfrac{63m}{\text{____}}$ Divide each side by ____.

____ $= m$ Simplify.

FOLDABLES

ORGANIZE IT

Write a proportion and find its cross products under the tab for Solve Proportions.

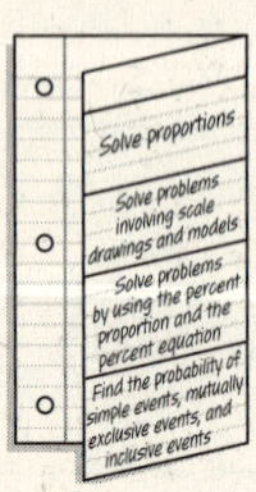

2 $\dfrac{z - 9}{4} = \dfrac{z}{5}$

$$\frac{z - 9}{4} = \frac{z}{5}$$

____$(z - 9) = 4\,$____ Find the cross products.

$5z -$ ____ $= 4z$ Distributive Property

$5z - 45 -$ ____ $= 4z -$ ____ Subtract ____ from each side.

$-45 = -z$

$\dfrac{-45}{\text{____}} = \dfrac{-z}{\text{____}}$ Divide each side by ____.

____ $= z$ Simplify.

3 **Convert 15 pints to quarts.**

Recall that 2 pints = 1 quart. Let x represent the number of quarts. Write a proportion.

$$\frac{15 \text{ pints}}{x \text{ quarts}} = \frac{2 \text{ pints}}{1 \text{ quart}}$$

$$\frac{15}{x} = \frac{2}{1}$$

$\square(1) = 2\square$ Find the cross products.

$\square = 2x$

$\frac{15}{\square} = \frac{2x}{\square}$ Divide each side by 2.

$\square = x$ So, 15 pints = $\square$ quarts.

REMEMBER IT

The denominator in a proportion cannot be equal to zero. Division by zero is undefined.

Your Turn **Solve each proportion.**

a. $\frac{26}{4} = \frac{n}{6}$

b. $\frac{b + 2}{b - 5} = \frac{9}{2}$

c. Convert 3 pounds to ounces.

BUILD YOUR VOCABULARY (pages 94–95)

The ______ of two measurements having different units of measure is called a **rate**.

A simplified ______ with a denominator of ______ is a **unit rate**.

The process of carrying units throughout a ______ is **dimensional analysis**.

EXAMPLE

4 **The density of copper is 8.96 grams per cubic centimeter. Suppose you have a piece of copper whose volume is 25 cubic centimeters. How many grams of copper do you have?**

8.96 grams per cubic centimeter = $\frac{8.96 \text{ grams}}{1 \text{ cubic centimeter}}$

Multiply the unit rate by the number of cubic centimeters of copper.

$\frac{\square \text{ grams}}{1} \cdot \frac{\square}{1} = \frac{\square}{1}$ or $\square$ grams

So, the piece of copper contains $\square$ grams of copper.

Your Turn The density of wood is 0.71 gram per cubic centimeter. Suppose you have a piece of wood whose volume is 60 cubic centimeters. How many grams of wood does it contain?

EXAMPLE

5 **A trucker drove 210 miles in 5 hours. At this rate, how far will she travel in 8 hours?**

Write the rate 210 miles in 5 hours as a unit rate. Then multiply by 8.

$\frac{210 \text{ miles}}{5 \text{ hours}} = \frac{42 \text{ miles}}{1 \text{ hour}}$ The unit rate is 42 miles per hour.

$8 \text{ hours} \times \frac{42 \text{ miles}}{1 \text{ hour}} = \frac{8}{1} \times \frac{42 \text{ miles}}{1}$ Note that the units cancel.

$= \square$ miles

Your Turn Damien drove 220 miles in 4 hours. At this rate, how far will he travel in 6 hours?

HOMEWORK ASSIGNMENT

Page(s):

Exercises:

5–2 Scale Drawing and Models

BUILD YOUR VOCABULARY (page 95)

A **scale drawing** or **scale model** is used to represent an object that is too ______ or too ______ to be drawn or built at actual size.

A **scale** is the ______ of the length of a model to the corresponding length of the ______ object.

WHAT YOU'LL LEARN

- Solve problems involving scale drawings and models.

EXAMPLE

1 The scale on a map of the upper Midwest is 1 inch = 15 miles. Find the distance between Chicago and Milwaukee on the map if the distance between the two cities is 90 miles.

Use the ______ and the distance between the cities to write a ______.

$$\frac{1 \text{ inch}}{\_\_\_\_\_\_} = \frac{x \text{ inches}}{\_\_\_\_\_\_}$$

← map distance

← actual distance

$1(\_\_\_\_\_\_) = 15(\_\_\_\_\_\_)$ Find the cross products.

$\_\_\_\_\_\_ = 15x$

$\frac{90}{\_\_\_\_\_\_} = \frac{15x}{\_\_\_\_\_\_}$ Divide each side by ______.

$\_\_\_\_\_\_ = x$

The distance between Chicago and Milwaukee on the map is about ______.

FOLDABLES™ ORGANIZE IT

Give examples of scale drawings or models under the tab for Solve Problems involving Scale Drawings and Models.

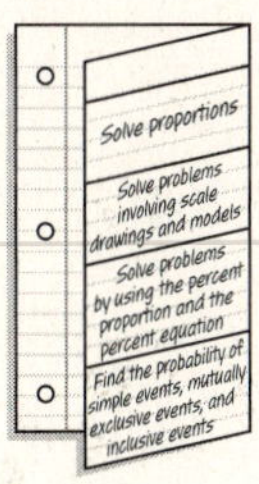

Your Turn The scale on a map of Alaska is $\frac{7}{8}$ inch = 150 miles. Find the actual distance between Fairbanks and Barrow if the distance between them on the map is $3\frac{1}{8}$ inches.

EXAMPLE

2 A railroad car is 36 feet long and a scale model of the railroad car is 1.5 feet long. What is the scale for the model?

Write the ratio of the length of the model to the length of the railroad car. Then solve a proportion in which the length of the model is 1 foot and the length of the railroad car is x feet.

$$\frac{1.5 \text{ feet}}{\square} = \frac{1 \text{ foot}}{x \text{ feet}}$$ ← model length ← actual length

$1.5x = \square(1)$ Find the $\square$ products.

$1.5x = \square$

$\frac{1.5x}{\square} = \frac{36}{\square}$ Divide each side by $\square$.

$x = \square$

The scale is 1 foot = $\square$ feet or $\square$.

Your Turn The height of the Statue of Liberty, from the heels to the top of the head, is about 111 feet. If a model of the statue is 5.5 feet from the heel to the top of the head, find the scale of the model.

HOMEWORK ASSIGNMENT

Page(s):

Exercises:

5–3 The Percent Proportion

BUILD YOUR VOCABULARY (page 95)

WHAT YOU'LL LEARN

- Solve problems by using the percent proportion.

A **percent** is a ______ that compares a number to ______.

The number that is divided into the ______ is the **base**.

The number that is divided by the ______ is the **percentage**.

EXAMPLE

1 Express the fraction as a percent.
$\frac{5}{8}$ of the square is shaded.

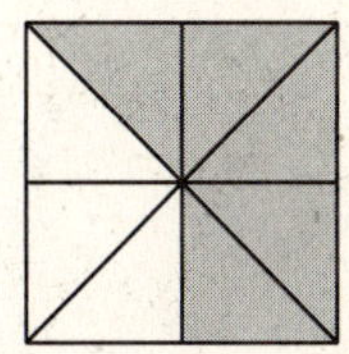

$\frac{5}{8} = \frac{r}{100}$ — r is the percent.

5 ______ $= 8$ ______ — Find the cross products.

______ $= 8r$

$\frac{500}{8} = \frac{8r}{8}$ — Divide each side by 8.

______ $= r$ — So, $\frac{5}{8} =$ ______.

Your Turn **Express each fraction or ratio as a percent.**

a. $\frac{4}{5}$ of the circle is shaded.

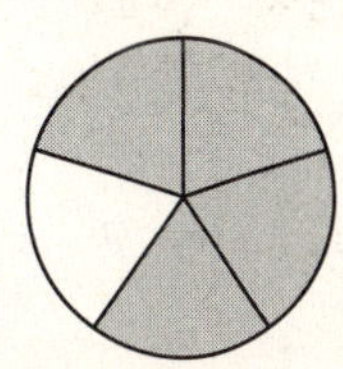

______

b. On Friday, 12 out of 60 students brought a sack lunch to school.

______

EXAMPLES

2 What percent of 175 is 35?

$\frac{P}{B} = \frac{r}{100}$ Use the percent proportion.

$\frac{\square}{\square} = \frac{r}{100}$ Replace P with ☐ and B with ☐.

$\square(100) = \square r$ Find the cross products.

$3500 = 175r$

$\frac{3500}{175} = \frac{175r}{175}$ Divide each side by 175.

$\square = r$ So, ☐ of 175 is 35.

KEY CONCEPT

Percent Proportion
If P is the percentage, B is the base, and r is the percent, the percent proportion is $\frac{P}{B} = \frac{r}{100}$.

3 20 is 40% of what number?

$\frac{P}{B} = \frac{r}{100}$ Use the percent proportion.

$\frac{20}{B} = \frac{40}{100}$ Replace P with 20 and r with 40.

$\square(100) = \square B$ Find the ☐ products.

$\square = 40B$

$\frac{2000}{\square} = \frac{40r}{\square}$ Divide each side by ☐.

$\square = r$ So, 20 is ☐ of 50.

Your Turn

a. What percent of 80 is 60?

b. 25% of what number is 67.5?

BUILD YOUR VOCABULARY (page 94)

A **circle graph** shows the relationship between parts and the whole.

EXAMPLE

4 A family recently moved to a new home. The table shows how much time the family spent packing boxes, cleaning the two homes, and unpacking boxes. What percent of the time did the family spend on each activity?

Activity	Time (days)
packing	7
cleaning	5
unpacking	8

The family worked for 7 + 5 + 8 or 20 days. This is the base. To find each percent, write and solve the percent proportion for each activity.

Packing: $\frac{\square}{\square} = \frac{r}{100}$ 7 [×] 100 [÷] 20 [=] ☐

Cleaning: $\frac{\square}{\square} = \frac{r}{100}$ 5 [×] 100 [÷] 20 [=] ☐

Unpacking: $\frac{\square}{\square} = \frac{r}{100}$ 8 [×] 100 [÷] 20 [=] ☐

The family spent ☐ of the time packing, ☐ of the time cleaning, and ☐ of the time unpacking.

Your Turn The table shows the number of hours Timothy spent on three activities. What percent of the time did Timothy spend on each activity?

Activity	Time (h)
reading	1
sports	4
homework	3

HOMEWORK ASSIGNMENT

Page(s):

Exercises:

5–4 The Percent Equation

WHAT YOU'LL LEARN

- Solve problems by using the percent equation.

KEY CONCEPT

Percent Equation
The percentage is equal to the rate times the base.

FOLDABLES Write the formulas for percent proportion and the percent equation under the tab for Solve Problems by Using the Percent Proportion and the Percent Equation.

EXAMPLES

1 Find 17% of $250.

$P = RB$ — Use the percent equation.

$= \square(\square)$ — Replace R with ______ and B with ______.

$= \square$ — 0.17 [×] 250 [ENTER] ______

So, 17% of $250 is ______.

2 35% of what number is 105?

$P = RB$ — Use the percent equation.

$\square = \square B$ — Replace P with ______ and R with ______.

$\frac{105}{\square} = \frac{0.35B}{\square}$ — Divide each side by ______.

$\square = B$ — 105 [÷] 0.35 [ENTER] ______

So, 35% of ______ is 105.

Your Turn

a. Find 12% of 360.

b. 19 is 25% of what number?

EXAMPLE

3 Riona serves food at a restaurant where she is paid 18% of the diners' bills. She earned $126 last weekend. What was the total of the diners' bills?

$P = RB$ Use the percent equation.

_____ = _____B Replace P with _____ and R with _____.

$\frac{126}{\_\_\_\_\_} = \frac{0.18B}{\_\_\_\_\_}$ Divide each side by _____.

_____ $= B$ 126 [÷] 0.18 [ENTER] _____

The total of the diners' bills was _____.

Your Turn A restaurant collects 7% sales tax on all items sold. If $35 is collected in one day, what are the total sales for that day?

BUILD YOUR VOCABULARY (pages 94–95)

Mixture problems involve combining _____ or more parts into a _____. The parts that are combined usually have a different price or a different percent of something.

EXAMPLE

4 **All 208 freshmen at a school went on a field trip. For transportation, buses that each hold 64 students and vans that each hold 8 students were used. Every bus and van was completely filled, and there were 5 vehicles used. How many buses and vans were used?**

	Number of Vehicles	Capacity	Total Capacity
Buses	b	64	$64b$
Vans	$5 - b$	8	$8(5 - b)$

students on buses + students on vans = total students

[ ] + [ ] = [ ]

$64b +$ [ ] $-$ [ ] $= 208$ Distributive Property

[ ] $+ 40 = 208$ $64b - 8b =$ [ ]

$56b =$ [ ]

$\frac{56b}{[\ ]} = \frac{168}{[\ ]}$ Divide.

$b =$ [ ]

There were [ ] buses and [ ] $-$ [ ] or [ ] vans.

Your Turn A T-shirt shop sells adult shirts for $15 each and children's shirts for $8 each. If $324 was collected for 30 shirts, how many of each type were sold?

REVIEW IT

Explain why 40 is subtracted from each side before each side is divided by 56 in Example 4. *(Lesson 4-5)*

HOMEWORK ASSIGNMENT

Page(s):

Exercises:

5–5 Percent of Change

WHAT YOU'LL LEARN

- Solve problems involving percent of increase and decrease.

BUILD YOUR VOCABULARY (page 95)

When an increase or decrease is expressed as a ________, the percent is called the **percent of increase** or **percent of decrease**.

EXAMPLE

1 Find the percent of increase or decrease. Round to the nearest percent.

original: 110

new: 140

This is an increase. The amount of increase is

________ – ________ or ________. Use the percent proportion.

$$\frac{P}{B} = \frac{r}{100}$$

$$\frac{\square}{\square} = \frac{r}{100}$$

$$\square(100) = \square r \qquad \text{Cross products}$$

$$\square = 110r$$

$$\frac{3000}{100} = \frac{110r}{110} \qquad \text{Divide each side by 110.}$$

$$\square \approx r$$

The percent of increase is about ________.

REMEMBER IT

Replace the equal sign with ≈ to indicate that your answer is approximate.

Your Turn Find the percent of increase or decrease. Round to the nearest percent. original: 180, new: 153

____

EXAMPLES

REMEMBER IT

There may be more than one way to solve a problem. Refer to page 213 in your textbook for another way to solve Examples 2 and 3.

2 A family bought a home computer for $890. A sales tax of 5.5% on the purchase was then added. What was the total price?

First, use the percent equation to find the sales tax.

$P = RB$

= ______ ______

= 48.95

Then add ______ to $890.

890 + ______ = ______

The total price was ______.

3 Suppose the family also bought a printer. The original price of the printer was $140, but they received a 15% discount. What was the sale price of the printer (before the sales tax was added)?

A discount of 15% means that the family will pay ______ − 15% or ______ of the price of the printer.

Use the percent equation to find the sale price.

$P = RB$

= ______(140)

= ______

The sale price of the printer is ______.

Your Turn

a. What is the total cost of a dress that sells for $60 if the sales tax rate is 5%?

______

b. All hamsters are on sale for 20% off. What is the sale price of a hamster that normally sells for $9.95?

______

HOMEWORK ASSIGNMENT

Page(s):

Exercises:

5-6 Probability and Odds

BUILD YOUR VOCABULARY (pages 94–95)

When all possible _______ have an equally likely chance of happening, the outcomes are said to be **random.**

Theoretical probability is what should occur.

What actually occurs when conducting an _______ is called the **experimental probability**.

WHAT YOU'LL LEARN

- Find the probability and odds of a simple event.

KEY CONCEPT

Probability The probability of an event is a ratio that compares the number of favorable outcomes to the number of possible outcomes.

EXAMPLE

1 Use the graph shown. If a person is chosen at random, what is the probability that the person is age 5–17?

Population of California

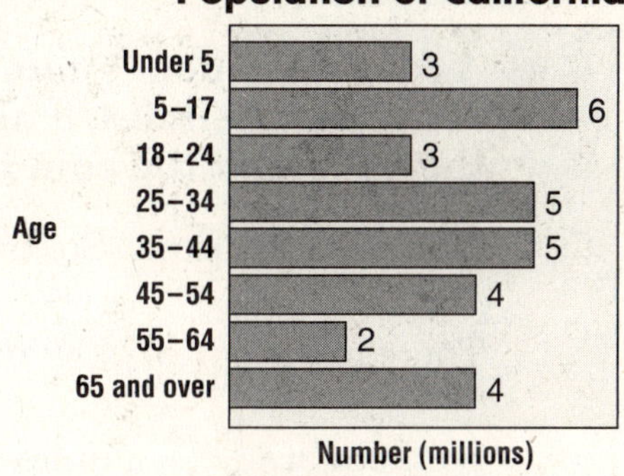

Source: U.S. Census Bureau

There are _______ million people who are between the ages of 5 and 17. The total population is _______ million.

$$P\text{ (age 5–17)} = \frac{\text{number of people age 5–17}}{\text{total population}}$$

$$= \frac{\square}{\square} \text{ or } \frac{3}{\square}$$

The probability is _______ or _______.

Your Turn A bag contains 1 yellow crayon, 3 red crayons, 4 blue crayons, and 7 green crayons. Suppose a crayon is chosen at random. What is the probability that it is yellow?

BUILD YOUR VOCABULARY (page 94)

A ______ that compares the number of ways an event can occur to the number of ______ it cannot occur is **odds**.

EXAMPLE

2 A coin is randomly removed from a change purse that contains 7 pennies, 8 nickels, and 5 quarters. What are the odds that the coin is a nickel?

There are 8 nickels. So, there are 8 favorable outcomes.

There are ______ + ______ or ______ coins that are not nickels.

So there are ______ unfavorable outcomes.

odds of choosing a nickel = ______ or ______

Your Turn A bag contains 1 yellow crayon, 3 red crayons, 4 blue crayons, and 7 green crayons. Find the odds of choosing a green crayon.

HOMEWORK ASSIGNMENT

Page(s):

Exercises:

5–7 Compound Events

What You'll Learn

- Find the probability of mutually exclusive and inclusive events.

Build Your Vocabulary (page 94)

Two or more simple events connected by the words ____ or ____ are called **compound events**.

The outcome of one event does not affect the outcome of the other event when they are **independent events**.

Two events that cannot occur at the same time are **mutually exclusive**.

Two events that can occur at the ____ time are **inclusive**.

Example

1 The two spinners are spun. Find the probability that the left spinner lands on green and the right spinner lands on a number greater than 2.

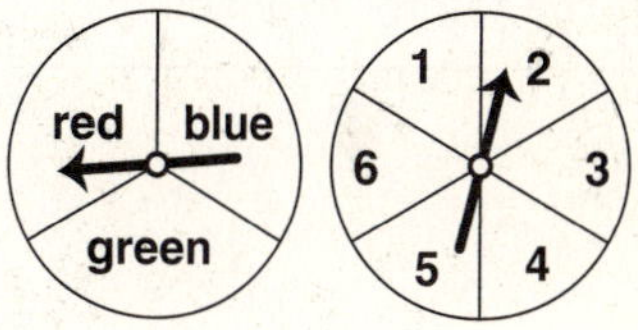

$P(\text{green}) = \frac{\square}{3}$ $\qquad (P > 2) = \frac{\square}{6}$ or $\frac{\square}{3}$

$P(\text{green and} > 2) = \frac{\square}{3} \cdot \frac{\square}{3}$ or $\frac{\square}{9}$

Your Turn Two dice are rolled. Find the probability that an even number is rolled on the first die and a number less than 3 is rolled on the second.

Key Concepts

Probability of Independent Events
The probability of two independent events is found by multiplying the probability of the first event by the probability of the second event.

Probability of Mutually Exclusive Events
The probability of two mutally exclusive events is found by adding the probability of the first event and the probability of the second event.

EXAMPLES

2 **A marble is selected at random from a bag that contains 5 red marbles, 3 blue marbles, and 2 yellow marbles. What is the probability that the marble is either red or yellow?**

A marble cannot be both red and yellow, so the events are mutually exclusive. Find the sum of the individual probabilities.

P (red or yellow) = P (red) + P (yellow)

$$= \frac{\square}{10} + \frac{\square}{10} \text{ or } \frac{\square}{10}$$

Key Concept

Probability of Inclusive Events The probability of two inclusive events is found by adding the probabilities of the events, then subtracting the probability of both events.

FOLDABLES Explain compound events, independent events and inclusive events, under the tab for probability.

3 **If there is a 90% chance of snow in January and a 95% chance of snow in February, find the probability that it will snow sometime in January or February.**

Since it is possible to snow in both months, these events are inclusive.

P (January) = ______ P (February) = ______

These events are independent since the weather in January does not affect the weather in February.

P (January *or* February)

= P (January) + P (February) − P(January and February)

= 0.9 + 0.95 − ______

= 1.85 − ______

= ______ or ______

Your Turn

a. Refer to the spinners in Example 1. Find the probability that the left spinner lands on red or the right spinner lands on 3.

______

b. If there is a 50% chance of rain on Monday and a 20% chance of rain on Tuesday, find the probability that it will rain sometime on Monday or Tuesday.

______

Homework Assignment

Page(s): ______

Exercises: ______

CHAPTER 5

BRINGING IT ALL TOGETHER

STUDY GUIDE

FOLDABLES™	VOCABULARY PUZZLEMAKER	BUILD YOUR VOCABULARY
Use your **Chapter 5 Foldable** to help you study for your chapter test.	To make a crossword puzzle, word search, or jumble puzzle of the vocabulary words in Chapter 5, go to: www.glencoe.com/sec/math/t_resources/free/index.php	You can use your completed **Vocabulary Builder** (pages 94–95) to help you solve the puzzle.

5-1 Solving Proportions

For each proportion, write the cross products.

1. $\frac{14}{35} = \frac{6}{15}$

2. $\frac{6}{8} = \frac{12}{16}$

3. A jet flying at a steady speed traveled 825 miles in 2 hours. How far did the jet travel in 1.5 hours?

5-2 Scale Drawings and Models

4. A model car is 12 centimeters long. The real car is 12 feet long. What is the scale of the model?

Write a proportion and solve for each of the following.

5. A 24-inch tall model was made in a scale of 1:3. What is the height of the actual object?

6. A flower that is 18 inches long is drawn to a scale of 1 centimeter to 1 inch. What is the height of the flower in the drawing?

5-3

The Percent Proportion

Use the percent proportion to find each number.

7. 12 is what percent of 36?

8. 16 is 20% of what number?

9. 75% of 28 is what number?

10. The table shows how Lavonne spends her day. What percent of the time does she spend on each activity? Round each answer to the nearest percent.

Activity	Time (hr)
sleep	8
school	6
work	3
homework	2
other	5

11. Explain how to check that the percentages are correct.

5-4

The Percent Equation

Use the percent equation to find each number.

12. 25 is 30% of what number?

13. What is 15% of 200?

14. Find 25% of 15.

5-5

Percent of Change

Find the percent of increase or decrease. Round to the nearest percent.

15. original: 50
new: 42

16. original: 50
new: 58

17. When you find a discount price, do you add to or subtract from the original price?

5-6

Probability and Odds

Write whether each statement is *true* or *false*. If false, replace the underlined word or number to make a true statement.

18. <u>Probability</u> can be written as a fraction, a decimal, or a percent.

19. The outcomes happen at <u>random</u> when all outcomes are equally likely to happen.

20. The probability of an impossible event is <u>1</u>.

21. The odds against an event occurring are the odds that the event <u>will</u> occur.

5-7

Compound Events

22. Two dice are rolled. Find the probability that an even number is rolled on the first die and the number 5 is rolled on the second die.

23. A sock contains 2 red marbles, 2 yellow marbles, and 6 blue marbles. One marble is chosen at random. What is the probability that the marble is either red or yellow?

CHAPTER 5

Checklist

ARE YOU READY FOR THE CHAPTER TEST?

Math Online

Visit **algconcepts.com** to access your textbook, more examples, self-check quizzes, and practice tests to help you study the concepts in Chapter 5.

Check the one that applies. Suggestions to help you study are given with each item.

☐ **I completed the review of all or most lessons without using my notes or asking for help.**

- You are probably ready for the Chapter Test.
- You may want take the Chapter 5 Practice Test on page 233 of your textbook as a final check.

☐ **I used my Foldable or Study Notebook to complete the review of all or most lessons.**

- You should complete the Chapter 5 Study Guide and Review on pages 230–232 of your textbook.
- If you are unsure of any concepts or skills, refer back to the specific lesson(s).
- You may also want to take the Chapter 5 Practice Test on page 233.

☐ **I asked for help from someone else to complete the review of all or most lessons.**

- You should review the examples and concepts in your Study Notebook and Chapter 5 Foldable.
- Then complete the Chapter 5 Study Guide and Review on pages 230–232 of your textbook.
- If you are unsure of any concepts or skills, refer back to the specific lesson(s).
- You may also want to take the Chapter 5 Practice Test on page 233.

Student Signature

Parent/Guardian Signature

Teacher Signature

CHAPTER 6

Functions and Graphs

FOLDABLES™ Use the instructions below to make a Foldable to help you organize your notes as you study the chapter. You will see Foldable reminders in the margin of this Interactive Study Notebook to help you in taking notes.

Begin with a sheet of notebook paper.

STEP 1 **Fold**
Fold lengthwise to the holes.

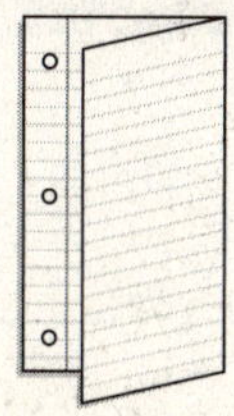

STEP 2 **Cut**
Cut along the top line and then cut 10 tabs.

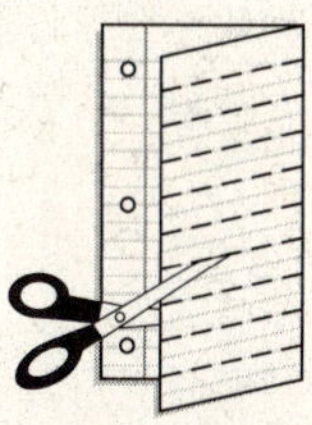

STEP 3 **Label**
Label the tabs using the vocabulary words as shown.

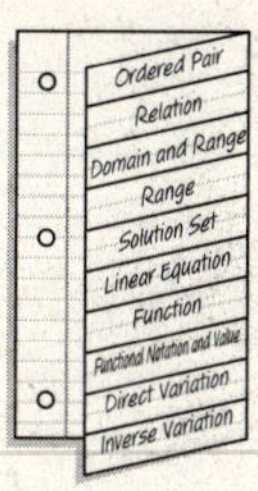

NOTE-TAKING TIP: When you take notes, it is helpful to write definitions and examples for each of the vocabulary terms.

Chapter 6

CHAPTER 6

BUILD YOUR VOCABULARY

This is an alphabetical list of new vocabulary terms you will learn in Chapter 6. As you complete the study notes for the chapter, you will see Build Your Vocabulary reminders to complete each term's definition or description on these pages. Remember to add the textbook page number in the second column for reference when you study.

Vocabulary Term	Found on Page	Definition	Description or Example
constant of variation [VARE-ee-AY-shun]			
dependent variable			
direct variation			
domain			
equation in two variables			
function			
functional notation			
functional value			
functional variable			

Vocabulary Term	Found on Page	Definition	Description or Example
independent variable			
inverse variation			
linear equation [LIN-ee-ur]			
range			
rate problem			
relation			
solution set			
vertical line test			
x-coordinate			
y-coordinate			

6–1 Relations

BUILD YOUR VOCABULARY (page 119)

The first number in an ordered pair is the **x-coordinate**.

The ______ number in an ordered pair is the ***y*-coordinate.**

A set of ______ pairs is a **relation**.

WHAT YOU'LL LEARN

- Show relations as sets of ordered pairs, as tables, and as graphs.

KEY CONCEPTS

Domain and Range of a Relation The domain of a relation is the set of all first coordinates from the ordered pairs of the relation.

The range of the relation is the set of all second coordinates from the ordered pairs of the relation.

FOLDABLES™ Write the definition for a relation and give an example under the tab for Relation.

EXAMPLE

1 Express the relation {(−4, 5), (−3, 2), (0, 1), (1, −1), (3, −2)} as a table and as a graph. Then determine the domain and range.

The domain is ______

and the range is ______

x	*y*
−4	5
0	1
3	−2

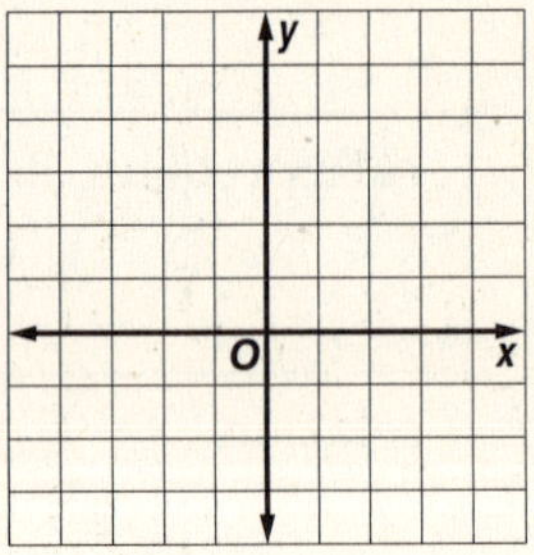

Your Turn Express the relation {(−5, 2), (−3, 1), (−2, −1), (0, −2), (2, −2)} as a table and as a graph. Then determine the domain and range.

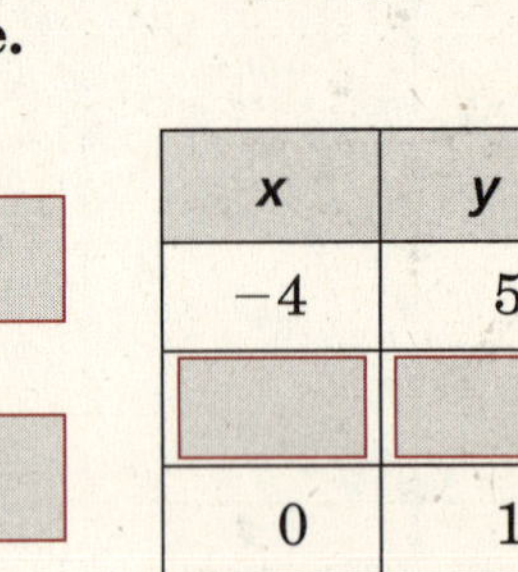

FOLDABLES™

ORGANIZE IT

Write a set of ordered pairs under the tab for Ordered Pair. Under the tab labeled Domain and Range, name the domain and range of the set of ordered pairs you chose.

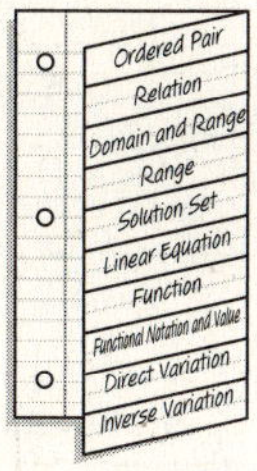

EXAMPLE

2 **Express the relation shown on the graph as a set of ordered pairs and in a table. Then find the domain and range.**

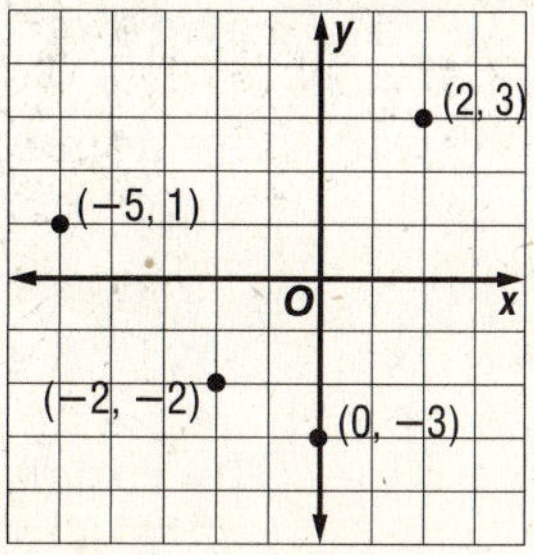

The set of ordered pairs for the relation is {(−5, 1), (   ), (   ), (, 3)}.

The domain is   , and the range is

x	y
−5	1
−2	−2
0	−3
2	3

Your Turn Express the relation shown on the graph as a set of ordered pairs and in a table. Find the domain and the range.

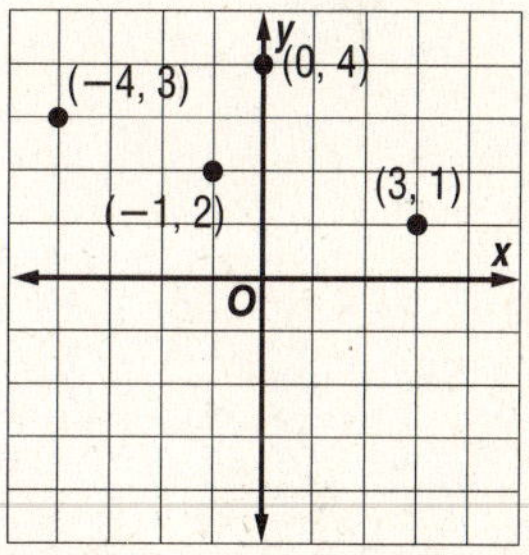

EXAMPLE

WRITE IT

Explain how you can tell which decade had the greatest increase in population.

______

______

______

______

______

______

3 The table shows the population of New York City since 1920.

Year	Population (millions)
1920	5.6
1930	6.9
1940	7.5
1950	7.9
1960	7.8
1970	7.9
1980	7.1
1990	7.3

a. Determine the domain and range of the relation.

The domain is

The range is

b. Graph the relation.

The x-coordinate goes from ______ to ______. The y-coordinates include values from from ______ to ______. You can include 0 and use units of 2.

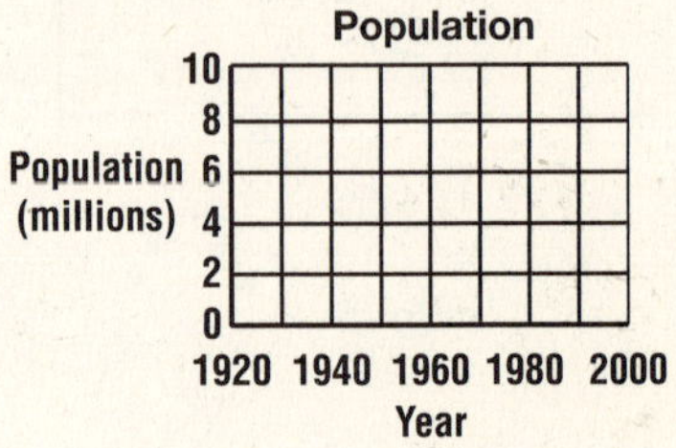

Your Turn **The table shows Colonial Population estimates from 1730 to 1780.**

Year	Population (in millions)
1730	0.6
1740	0.9
1750	1.2
1760	1.6
1770	2.1
1780	2.8

Source: www.infoplease.com

a. Determine the domain and range of the relation.

b. Graph the relation.

HOMEWORK ASSIGNMENT

Page(s):

Exercises:

6–2 Equations as Relations

WHAT YOU'LL LEARN

- Solve linear equations for a given domain.

BUILD YOUR VOCABULARY (pages 118–119)

A set of ______ to a problem is a **solution set.**

An equation that contains two ______ values is an **equation in two variables.**

KEY CONCEPT

Solution of an Equation in Two Variables If a true statement results when the numbers in an ordered pair are substituted into an equation in two variables, then the ordered pair is a solution of the equation.

FOLDABLES In your own words, tell what a solution set is under the tab for Solution Set.

EXAMPLE

1 Which of the ordered pairs (0, 0), (1, 4), (2, 1), or (−1, 2) are solutions of $y = -x + 3$?

Make a table. Substitute the x- and y-values of each ordered pair into the equation.

x	y	$y = -x + 3$	True or False?
0	0	$0 = (0) + 3$ ______ = ______	______
1	4	$4 = -(1) + 3$ ______ = ______	______
2	1	$1 = -(2) + 3$ ______ = ______	______
−1	2	$2 = -(-1) + 3$ ______ = ______	______

A ______ statement results when the ordered pair ______ is substituted into the equation. So, the ordered pair ______ is a solution of the equation $y = -x + 3$.

Your Turn Which of the ordered pairs (0, 2), (1, −3), (2, −5), or (−1, −2) are solutions of $y = -2x - 1$?

EXAMPLE

2 Solve $y = 2x + 1$ if the domain is {−2, −1, 0, 1, 2}. Graph the solution set.

Make a table. Substitute each value of ____ into the equation to determine the corresponding values of ____.

x	$2x + 1$	y	(x, y)
−2	2(−2) + 1		
−1	2(−1)+ 1		
0	2(0) + 1		
1	2(1) + 1		
2	2(2) + 1		

The solution set is ____.

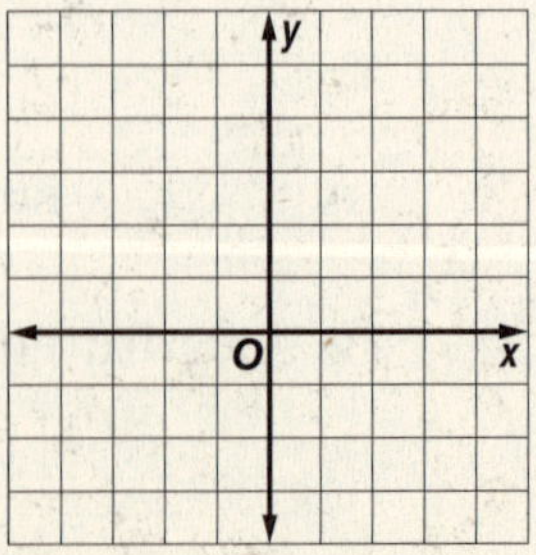

REMEMBER IT

Sometimes you can solve an equation for *y* before substituting each domain value into the equation. This makes creating a table of values easier.

Your Turn Solve $y = 3x - 2$ if the domain is $\{-2, -1, 0, 1, 2\}$. Graph the solution set.

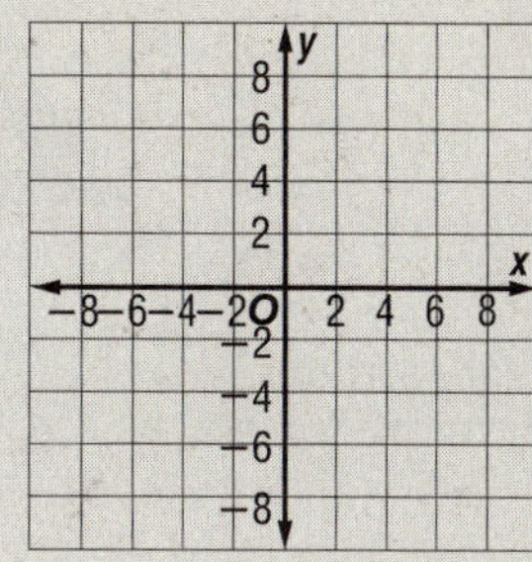

EXAMPLE

3 Find the domain of $y = 10 - 4x$ if the range is $\{-6, -2, 2, 6, 10\}$.

Make a table. Substitute each value of ______ into the equation. Then solve each equation to determine the corresponding values of ______.

y	$y = 10 - 4x$	x	(x, y)
−6	$-6 = 10 - 4x$		
−2	$-2 = 10 - 4x$		
2	$2 = 10 - 4x$		
6	$6 = 10 - 4x$		
10	$10 = 10 - 4x$		

The domain is ______.

Your Turn Find the domain of $y = 8 - 3x$ if the range is $\{-4, -1, 2, 5, 8\}$.

HOMEWORK ASSIGNMENT

Page(s):

Exercises:

6–3 Graphing Linear Relations

Build Your Vocabulary (page 119)

An equation with a graph that is a ______ is a **linear equation**.

What You'll Learn

- Graph linear relations.

Key Concept

Linear Equation in Standard Form A linear equation is an equation that can be written in the form $Ax + By = C$, where A, B, and C are any numbers, and A and B are not both zero. $Ax + By = C$ is called the *standard form* if A, B, and C are integers.

Foldables Write the standard form of a linear equation under the tab for Linear Equation. Be sure to include examples.

EXAMPLES

Determine whether each equation is a linear equation. Explain. If an equation is linear, identify *A*, *B*, and *C* when written in standard form.

1 $4xy = 4$

Since the term ______ has two variables, the equation cannot be written in the form ______.

So, this is not a ______.

2 $y = x$

First, rewrite the equation so that both variables are on the same side of the equation.

$y = x$

$0 = x - y$ Subtract ______ from each side.

This equation can be written as ______ Therefore, it is a linear equation in the form $Ax + By = C$, where

$A =$ ______, $B =$ ______, and $C =$ ______.

Your Turn **Determine whether each equation is a linear equation. Explain. If an equation is linear, identify *A*, *B*, and *C* when written in standard form.**

a. $2x = 3 + y$

______

b. $2x = 3xy$

______

EXAMPLE

3 **Graph $3x - y = 2$.**

In order to find values for y more easily, solve the equation for ______.

$3x - y = 2$

$-y = 2 -$ ______ Subtract ______ from each side.

$y = -2 +$ ______ Multiply each side by ______.

Now make a table and draw the graph.

x	$-2 + 3x$	y	(x, y)
−2	$-2 + 3(-2)$		
−1	$-2 + 3(-1)$		
0	$-2 + 3(0)$		
1	$-2 + 3(1)$		
2	$-2 + 3(2)$		

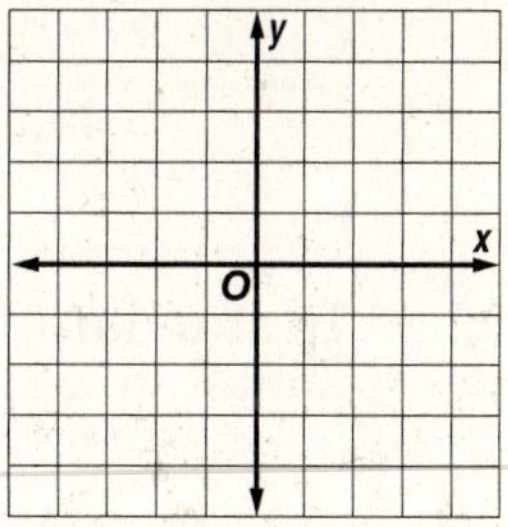

Your Turn **Graph each equation.**

a. $y = 2x - 2$

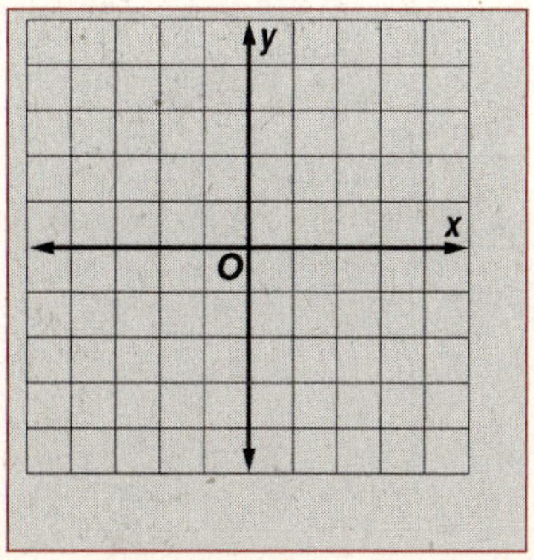

b. $x - y = -5$

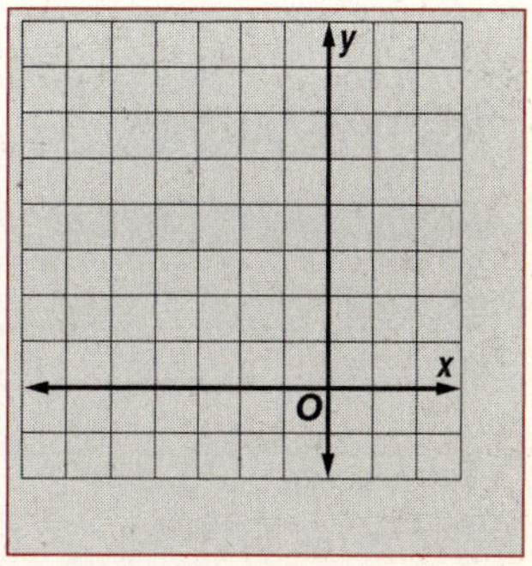

EXAMPLES

4 Graph $y = -2$.

In standard form, this equation is written as ____ $+ y = -2$. So, for any value of x, $y =$ ____. For example, if $x = 0$, $y =$ ____; if $x = 1$, $y =$ ____; if $x = 3$, $y =$ ____.

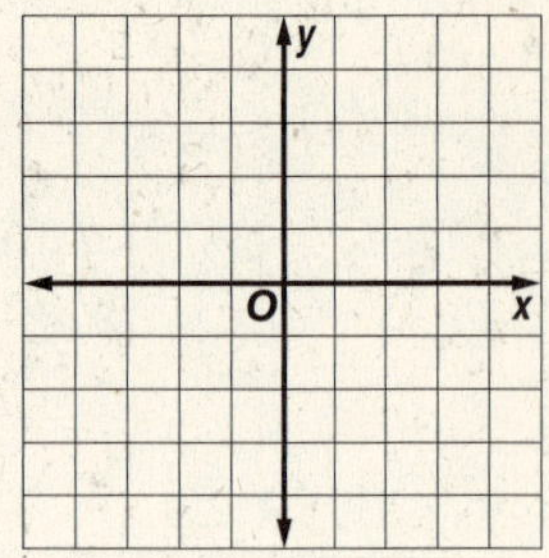

The graph of $y = -2$ is a horizontal line.

5 Graph $y = 3x$.

Make a table and draw the graph.

x	3x	y	(x, y)
−2	3(−2)		
−1	3(−1)		
0	3(0)		
1	3(1)		
2	3(2)		

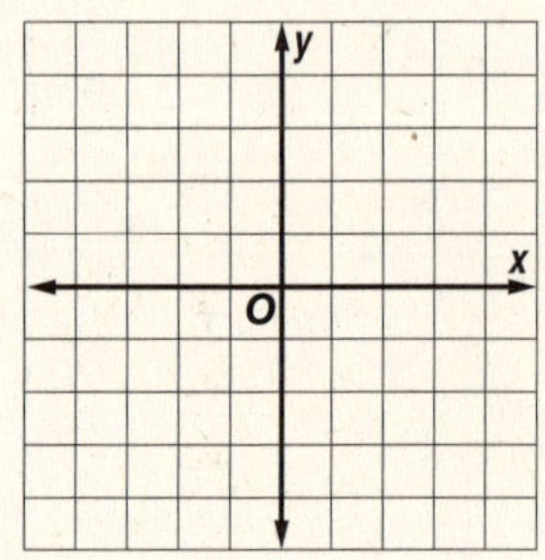

In standard form, $y = 3x$ is written as $3x - y = 0$. In any equation where $C = 0$, the graph passes through the origin.

Your Turn **Graph each equation.**

a. $y = -4$

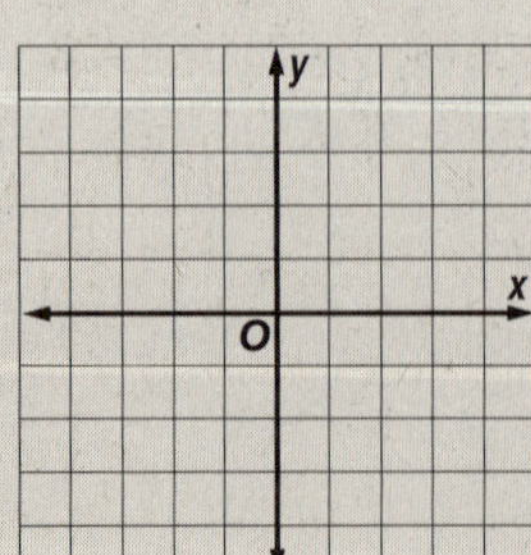

b. $y = -4x$

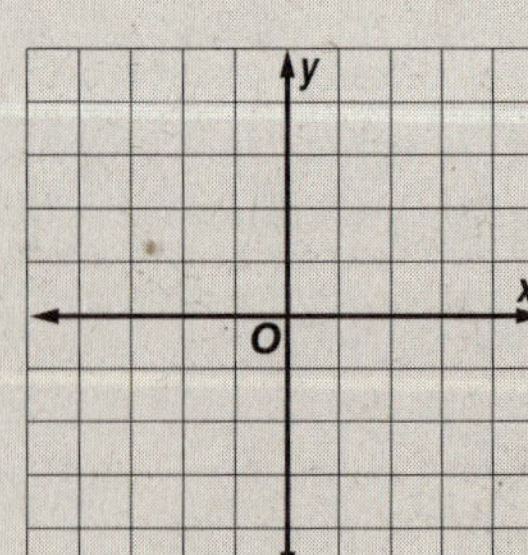

HOMEWORK ASSIGNMENT

Page(s):

Exercises:

6–4 Functions

EXAMPLES

What You'll Learn

- Determine whether a given relation is a function.

Key Concept

Function A function is a relation in which each member of the domain is paired with *exactly* one member of the range.

Foldables Write three examples of relations that are functions under the tab for Function. Write each function in a different form.

Determine whether each relation is a function. Explain your answer.

1 **{(1, 1), (3, 4), (4, 5), (4, 6)}**

This is not a function because one member of the domain, ____, is paired with two members of the range, ____ and ____.

2

x	−2	−1	0	1	2
y	0	1	2	3	4

The table represents a function since, for each member of the ____, there is only ____ corresponding member of the ____.

3

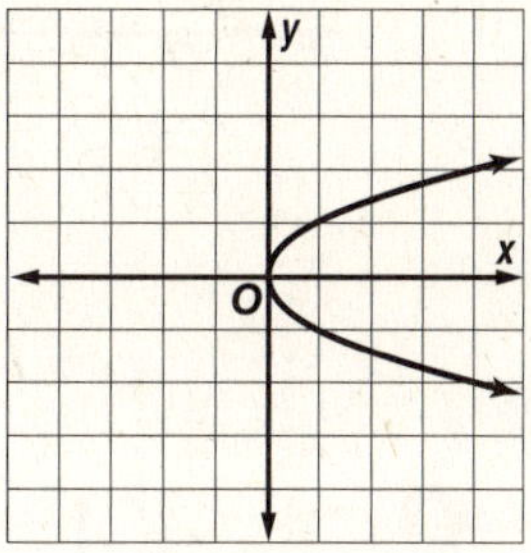

The graph represents a relation that is not a ____ because there are many members of the ____ (the x-values) that are paired with ____ members of the ____ (the y-values).

Your Turn **Determine whether each relation is a function. Explain your answer.**

a. {(1, 2), (2, 3), (3, 5), (4, 5)}

b.

x	−3	−1	1	1	2
y	−1	−2	−3	−4	−5

c.

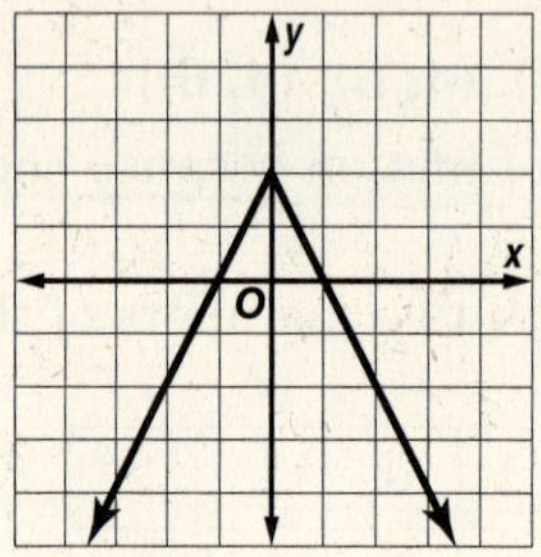

EXAMPLES

Key Concept

Vertical Line Test for a Function If each vertical line passes through no more than one point of the graph of a relation, then the relation is a function.

Use the vertical line test to determine whether each relation is a function.

4

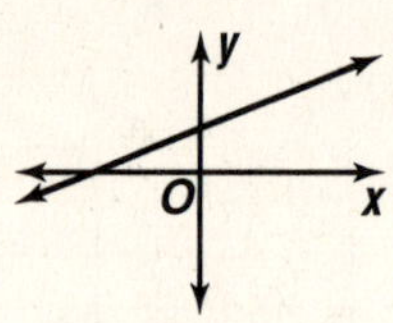

This relation is ______ since each vertical line passes through no more than ______ point of the ______ of the relation.

5

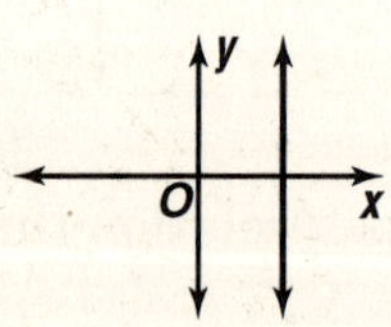

This relation is ______ since a ______ line passes through ______ point of the graph.

Your Turn **Use the vertical line test to determine whether each relation is a function.**

a.

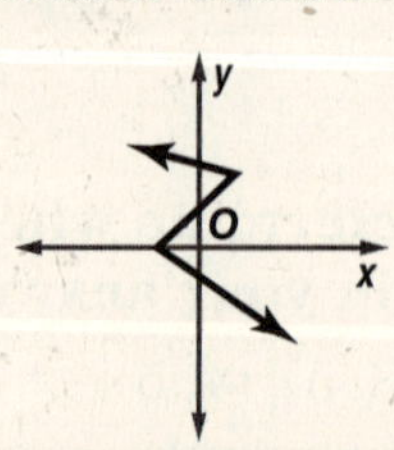

b.

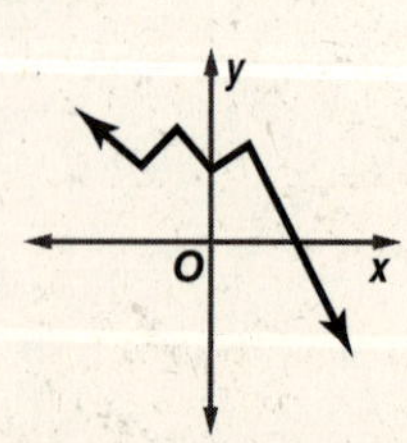

FOLDABLES™

ORGANIZE IT

Give an example of an equation in functional notation under the tab for Functional Notation and Value. Then choose a number and find the functional value of your function for the chosen number.

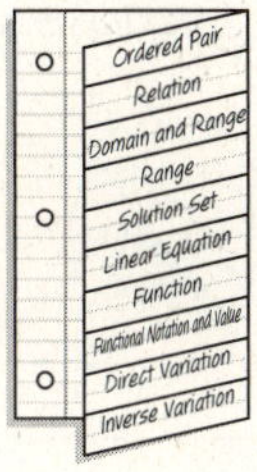

BUILD YOUR VOCABULARY (page 118)

Writing equations of the form "$y = \ldots$" as "$f(x) = \ldots$" is called **functional notation**.

A **functional value** is a ________ that corresponds to a specific ________.

EXAMPLES

If $f(x) = x - 4$, find each value.

6 $f(2)$

$f(2) = x - 4$

$f(2) =$ ____ $- 4$ Replace x with ____.

$=$ ____ Add.

7 $f\left(\frac{1}{2}\right)$

$f(x) = x - 4$

$f\left(\frac{1}{2}\right) =$ ____ $- 4$ Replace x with ____.

$=$ ____ Add.

8 $f(c)$

$f(x) = x - 4$

$f(c) =$ ____ $- 4$ Replace x with ____.

Your Turn **If $f(x) = -x + 4$, find each value.**

a. $f(-1)$ **b.** $f(1.5)$ **c.** $f(a)$

HOMEWORK ASSIGNMENT

Page(s):

Exercises:

6–5 Direct Variation

What You'll Learn

- Solve problems involving direct variations.

Build Your Vocabulary (pages 118–119)

The variable whose value depends on the value of the independent variable is the **dependent variable**.

The **independent variable** is the ______ whose value is chosen.

The **constant of variation** is the number ______ in an equation of the form $y = kx$.

Key Concept

Direct Variation A direct variation is a linear function that can be written in the form $y = kx$, where $k \neq 0$.

Foldables™ Write the standard form of an equation that is a direct variation under the tab for Direct Variation. Also draw the graph of a linear equation that varies directly.

Examples

Determine whether the equation is a direct variation.

1 $y = \frac{1}{2}x$

Graph the equation. The graph passes through the ______.

So, the equation is ______.

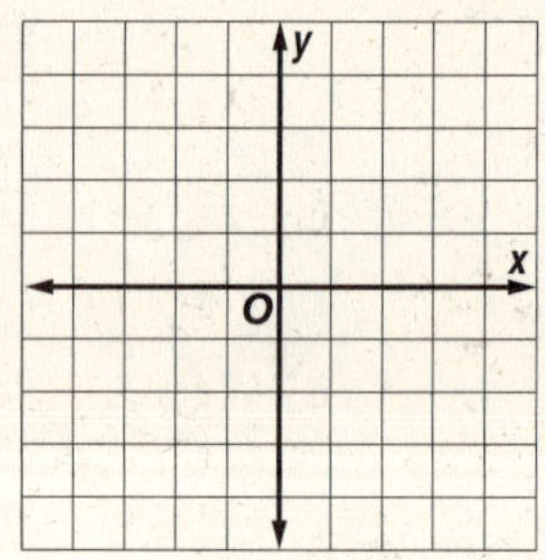

The constant of variation is ______.

2 $y = x + 2$

Graph the equation. The graph does not pass through the ______.

So, the equation is ______.

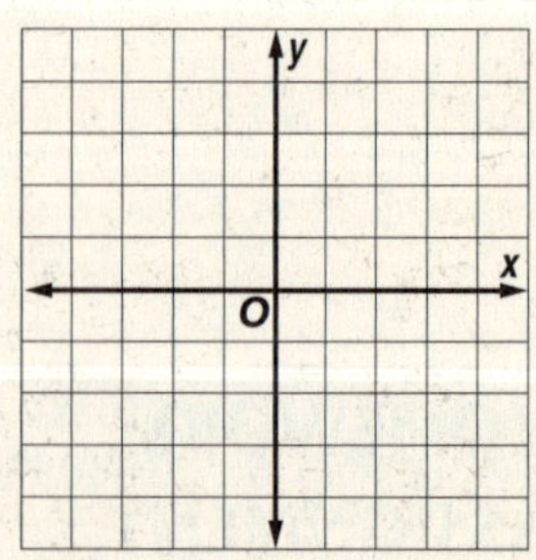

Your Turn **Determine whether each equation is a direct variation.**

a. $y = \frac{2}{3}x + 1$ ______

b. $y = -2x$ ______

BUILD YOUR VOCABULARY (page 119)

A problem involving the formula ______ = ______ × ______ or $d = rt$ is a **rate problem.**

EXAMPLE

3 The length of a trip varies directly as the amount of gasoline used. How many gallons of gasoline would be needed for a 550-mile trip if a 66-mile trip used 3 gallons of gasoline?

Let ℓ represent the length of the trip and let g represent the amount of gasoline used. The statement *the length varies directly as the amount of gasoline* translates into an equation $\ell = kg$ in the same way as *y varies directly as x* translates into $y = kx$. Find the value of k.

$\ell = kg$	Direct variation
______ $= k \cdot$ ______	Replace ℓ with ______ and g with ______.
______ = ______	Divide each side by ______.
______ $= k$	

Next, find the amount of gasoline needed for a 550-mile trip.

$\ell = kg$	
$550 =$ ______ $\times g$	Replace ℓ with 550 and k with ______.
$\frac{550}{\text{______}} = \frac{22 \cdot g}{\text{______}}$	Divide each side by ______.
______ $= g$	

A 550-mile trip would use ______ gallons of gasoline.

WRITE IT

In Example 3, explain the meaning of k within the context of the problem.

______

______

______

______

Your Turn How many gallons of gasoline would be needed for a 630-mile trip if a 126-mile trip used 7 gallons of gasoline?

EXAMPLE

Remember It

There are several combinations for proportions that can be used in direct variation. In addition to the proportion used in Example 4, $\frac{y_1}{y_2} = \frac{x_1}{x_2}$ and $\frac{x_1}{y_1} = \frac{x_2}{y_2}$ can also be used.

4 Suppose y varies directly as x and $y = 27$ when $x = 18$. Find x when $y = 15$.

Use $\frac{y_1}{x_1} = \frac{y_2}{x_2}$ to solve the problem.

$\frac{\square}{\square} = \frac{\square}{x_2}$ Let $y_1 = \square$, $y_2 = \square$, and $x_1 = \square$.

$\square x_2 = \square(15)$ Find the cross products.

$27x_2 = \square$

$\frac{27x_2}{\square} = \frac{270}{\square}$ Divide each side by $\square$.

$x_2 = \square$

So, $x = \square$ when $y = 15$.

Your Turn Suppose y varies directly as x and $y = 35$ when $x = 14$. Find x when $y = 15$.

Homework Assignment

Page(s):

Exercises:

6–6 Inverse Variation

What You'll Learn

- Solve problems involving inverse variations.

Key Concept

Inverse Variation
An inverse variation is described by an equation of the form $xy = k$, where $k \neq 0$.

EXAMPLE

1 The number of bricklayers needed to build a brick wall varies inversely as the number of hours needed. If 4 bricklayers can build a brick wall in 30 hours, how long would it take 5 bricklayers to do it?

Let x = the number of bricklayers. Let y = the number of hours. First find the value of k.

$xy = k$ Definition of inverse variation

$\square(\square) = k$ Replace x with $\square$ and y with $\square$.

$\square = k$ The constant of variation is $\square$.

Next, find the number of hours for 5 bricklayers to build a brick wall.

$y = \frac{k}{x}$ Divide each side of $xy = k$ by $\square$.

$= \frac{\square}{\square}$ Replace k with $\square$ and x with $\square$.

$= \square$

A crew of 5 bricklayers can build the wall in $\square$ hours.

Your Turn The number of painters needed to paint a bedroom varies inversely as the number of hours needed. If 3 painters can paint a bedroom in 8 hours, how long would it take 4 painters to do it?

EXAMPLE

2 **Suppose y varies as x and $y = -2$ when $x = -12$. Find y when $x = 8$.**

$\frac{x_1}{x_2} = \frac{y_2}{y_1}$ Inverse variation proportion

$\frac{\square}{\square} = \frac{y_2}{\square}$ Let x_1 = ____, y_1 = ____, and x_2 = ____.

$\square(-2) = \square y_2$ Find the cross products.

$\square = 8y_2$

$\square = y_2$

Therefore, when $x = 8$, y = ____.

REVIEW IT

Why does the product of −12 and −2 result in a positive answer? (*Lesson 2-5*)

____

____

____

____

Your Turn Suppose y varies inversely as x and $y = -5$ when $x = -9$. Find y when $x = 6$.

HOMEWORK ASSIGNMENT

Page(s):

Exercises:

CHAPTER 6

BRINGING IT ALL TOGETHER

STUDY GUIDE

FOLDABLES	VOCABULARY PUZZLEMAKER	BUILD YOUR VOCABULARY
Use your **Chapter 6 Foldable** to help you study for your chapter test.	To make a crossword puzzle, word search, or jumble puzzle of the vocabulary words in Chapter 6, go to: www.glencoe.com/sec/math/t_resources/free/index.php	You can use your completed **Vocabulary Builder** (pages 118–119) to help you solve the puzzle.

6-1

Relations

Match the relation on the left to its other form on the right.

1.

x	−1	1	0	2
y	3	2	−1	4

2. $\{(1, 5), (1, 3), (1, 0), (1, -2)\}$

3.

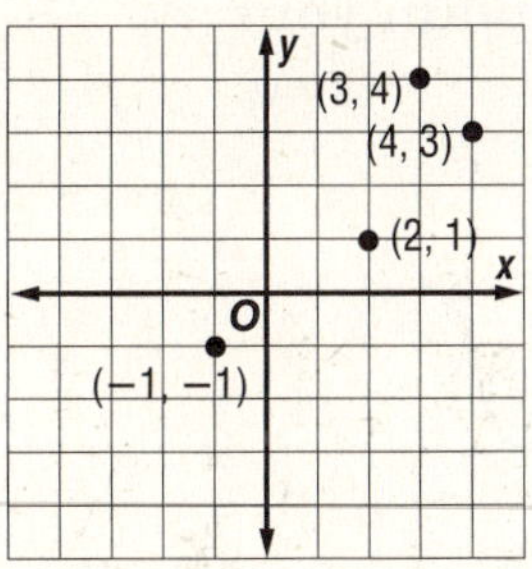

a.

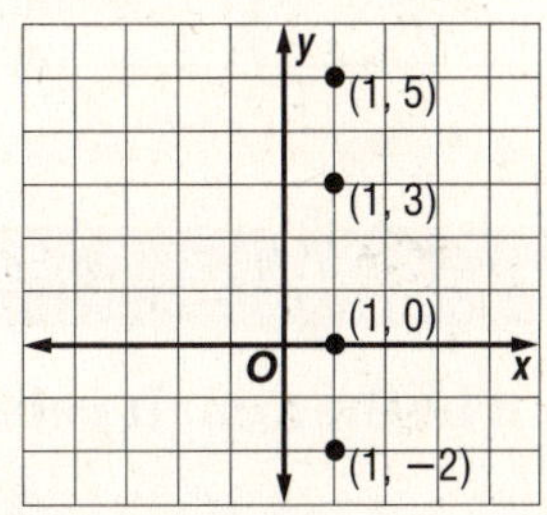

b. $\{(-1, 3), (1, 2), (0, -1), (2, 4)\}$

c.

x	4	1	3	−1
y	3	2	4	−1

d. $\{(3, 4), (2, 1), (4, 3), (-1, -1)\}$

4. Name the domain and range of the relation in Exercise 1.

Domain = Range =

6-2

Equations as Relations

Solve each equation if the domain is $\{-1, 0, 1, 2, 3\}$.

5. $y = 3x$

6. $y = 2 - x$

7. $x + y = 1$

6-3

Graphing Linear Relations

Complete the table.

	Equation	Linear or Nonlinear?	Reason
8.	$2x = 3y + 1$		
9.	$4xy + 2y = 7$		
10.	$2x^2 = 4y - 3$		
11.	$\frac{x}{5} - \frac{4y}{3} = 2$		

6-4

Functions

Determine whether each relation is a function. Explain your answer.

12.

x	−2	−1	0	4
y	1	2	3	7

13.

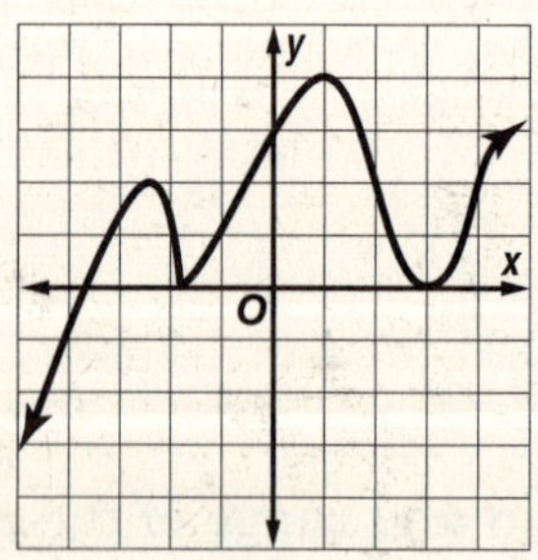

If $f(x) = 2x + 3$, find the following:

14. $f(0)$ **15.** $f(-2)$ **16.** $f(3)$ **17.** $f(b)$

6-5

Direct Variation

For each situation, write an equation with the proper constant of variation.

18. The distance d varies directly as time t.

19. The wages W earned by an employee vary directly with the number of hours h that are worked.

20. The length of a trip varies directly as the amount of gasoline used. Pedro's car used 4 gallons of gasoline in the first 112 miles of his trip. How much gasoline should he expect to use in the remaining 84 miles of the trip?

6-6

Inverse Variation

Write *direct variation*, *inverse variation*, or *neither* to describe the relationship between x and y described by each equation.

21. $y = 3x$ **22.** $xy = 5$ **23.** $y = 28x$

For each problem, y varies inversely as x. Write an equation you can use to solve for k. Then write a proportion you could use to solve the problem.

	Problem	Equation	Proportion
24.	If $y = 8$ when $x = 12$, find y when $x = 4$.		
25.	If $x = 50$ when $y = 6$, find x when $y = 30$.		

CHAPTER 6

Checklist

ARE YOU READY FOR THE CHAPTER TEST?

Math Online

Visit **algconcepts.com** to access your textbook, more examples, self-check quizzes, and practice tests to help you study the concepts in Chapter 6.

Check the one that applies. Suggestions to help you study are given with each item.

☐ **I completed the review of all or most lessons without using my notes or asking for help.**

- You are probably ready for the Chapter Test.
- You may want take the Chapter 6 Practice Test on page 279 of your textbook as a final check.

☐ **I used my Foldable or Study Notebook to complete the review of all or most lessons.**

- You should complete the Chapter 6 Study Guide and Review on pages 276–278 of your textbook.
- If you are unsure of any concepts or skills, refer back to the specific lesson(s).
- You may also want to take the Chapter 6 Practice Test on page 279.

☐ **I asked for help from someone else to complete the review of all or most lessons.**

- You should review the examples and concepts in your Study Notebook and Chapter 6 Foldable.
- Then complete the Chapter 6 Study Guide and Review on pages 276–278 of your textbook.
- If you are unsure of any concepts or skills, refer back to the specific lesson(s).
- You may also want to take the Chapter 6 Practice Test on page 279.

Student Signature

Parent/Guardian Signature

Teacher Signature

CHAPTER 7

Linear Equations

FOLDABLES™ Use the instructions below to make a Foldable to help you organize your notes as you study the chapter. You will see Foldable reminders in the margin of this Interactive Study Notebook to help you in taking notes.

Begin with four sheets of grid paper.

STEP 1 **Fold**
Fold each sheet in half from top to bottom.

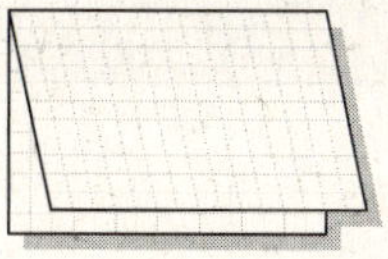

STEP 2 **Cut**
Cut along fold. Staple the eight half-sheets together to form a booklet.

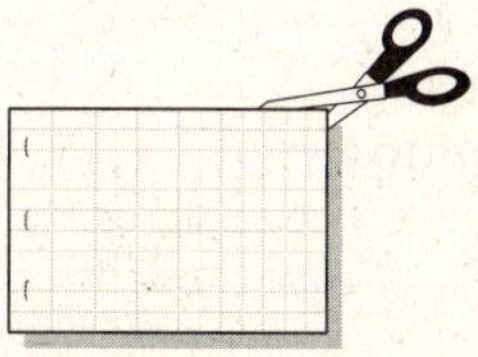

STEP 3 **Cut**
Cut tabs into margin. The top tab is 4 lines wide, the next tab is 8 lines wide, and so on.

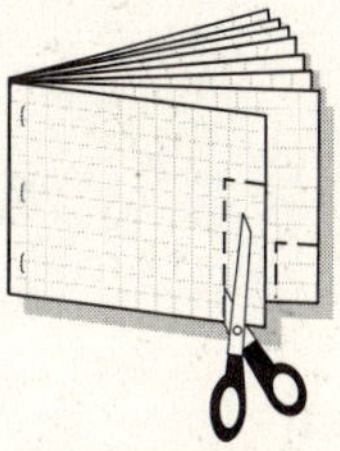

STEP 4 **Label**
Label each page with a lesson number and title. The last tab is for the vocabulary.

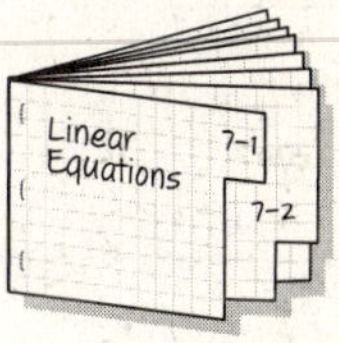

NOTE-TAKING TIP: When you take notes, don't lose focus as to what your teacher is saying. If you do, you will become lost in your note-taking. Write important equations and/or sketch graphs using the methods presented in each lesson.

CHAPTER 7

BUILD YOUR VOCABULARY

This is an alphabetical list of new vocabulary terms you will learn in Chapter 7. As you complete the study notes for the chapter, you will see Build Your Vocabulary reminders to complete each term's definition or description on these pages. Remember to add the textbook page number in the second column for reference when you study.

Vocabulary Term	Found on Page	Definition	Description or Example
best-fit line			
correlation coefficient [CORE-uh-LAY-shun]			
extrapolation [ek-STRA-puh-LAY-shun]			
family of graphs			
interpolation [in-TER-puh-LAY-shun]			
linear regression			
median-median line			
parallel lines [PARE-uh-lel]			
parent graph			

Vocabulary Term	Found on Page	Definition	Description or Example
perpendicular lines [PER-pun-DI-kyoo-lur]			
point-slope form			
rate of change			
residual			
rise			
run			
scatter plot			
slope			
slope-intercept form [IN-ter-SEPT]			
x-intercept			
y-intercept			

7-1 Slope

What You'll Learn

- Find the slope of a line given the coordinates of two points on the line.

Build Your Vocabulary (page 143)

Slope is the ratio of the ______, or the vertical change, to the ______, or the horizontal change.

Key Concept

Slope The slope of a line is the ratio of the change in *y* to the corresponding change in *x*.

Foldables Name two points and find the slope between them under the tab for Lesson 7-1. Then give examples of positive, negative, zero and undefined slopes.

Examples

Determine the slope of each line.

1

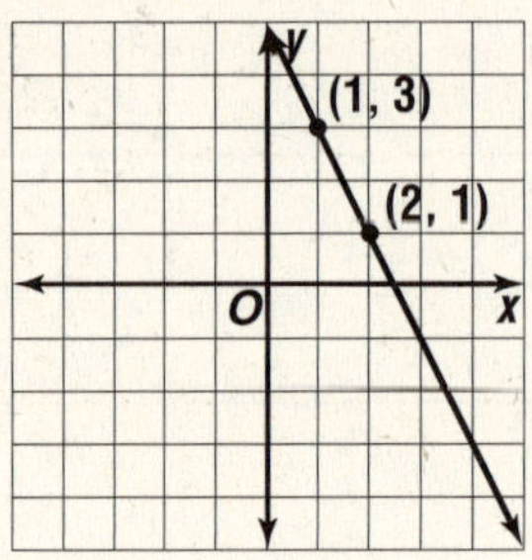

$$\text{slope} = \frac{\text{change in } y}{\text{change in } x}$$

$$= \frac{\square - 1}{\square - 2}$$

$$= \frac{\square}{-1}$$

The slope is ______.

2

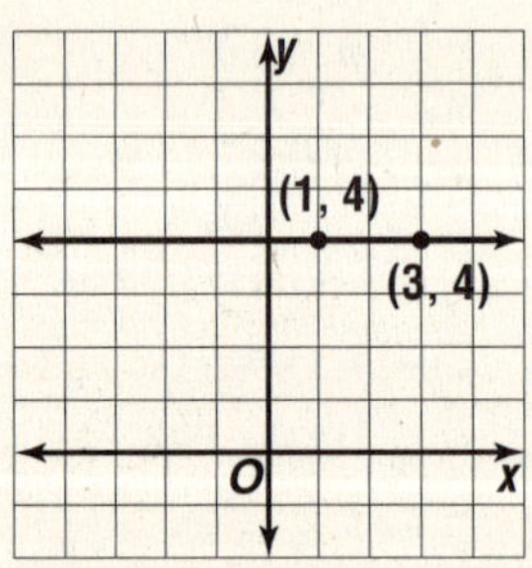

$$\text{slope} = \frac{\text{change in } y}{\text{change in } x}$$

$$= \frac{\square - 4}{\square - 3}$$

$$= \frac{\square}{-2}$$

The slope is ______.

Your Turn **Determine the slope of each line.**

a.

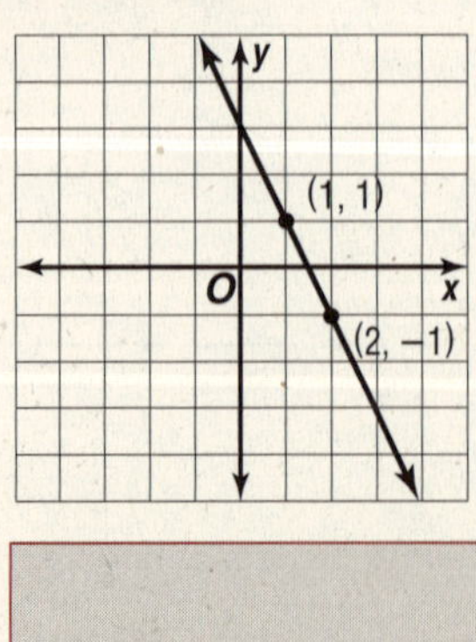

______

b.

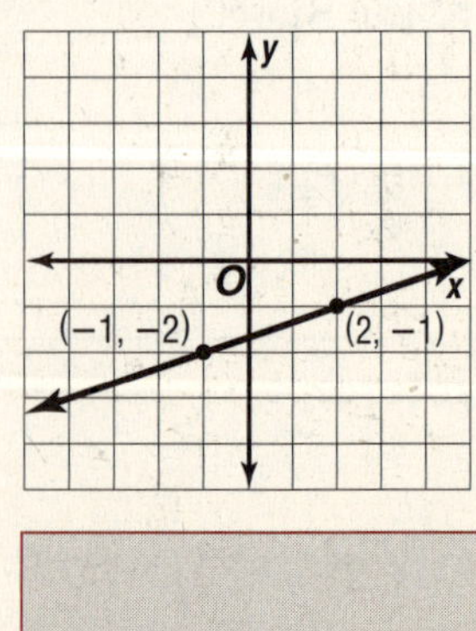

______

EXAMPLE

3 A line contains the points whose coordinates are listed in the table. Determine the slope of the line.

+1 +1 +1

x	−2	−1	0	1
y	7	4	1	−2

−3 −3 −3

Each time x increases ____ unit, y decreases ____ units.

$$\text{slope} = \frac{\text{change in } y}{\text{change in } x} = \frac{\_\_\_\_}{\_\_\_\_}$$

The slope of the line containing these points is ____.

Your Turn A line contains the points whose coordinates are listed in the table. Determine the slope of the line.

x	−3	−2	−1	0
y	2	4	6	8

KEY CONCEPT

Determining Slope Given Two Points The slope m of a line containing any two points (x_1, y_1) and (x_2, y_2) is given by the following formula.

$$\text{Slope} = \frac{y_2 - y_1}{x_2 - x_1}$$

EXAMPLE

4 Find the slope of a line that passes through (3, 8) and (3, 4).

$$m = \frac{y_2 - y_1}{x_2 - x_1}$$

$$= \frac{4 - \_\_\_\_}{3 - \_\_\_\_}$$

$$= \frac{\_\_\_\_}{0}$$

The slope is ____.

Your Turn Determine the slope of the line that passes through (5, 2) and (3, −1).

HOMEWORK ASSIGNMENT

Page(s):

Exercises:

7–2 Writing Equations in Point-Slope Form

WHAT YOU'LL LEARN

- Write a linear equation in point-slope form given the coordinates of a point on the line and the slope of the line.

KEY CONCEPT

Point-Slope Form For a nonvertical line through the point at (x_1, y_1) with slope m, the point-slope form of a linear equation is $y - y_1 = m(x - x_1)$

FOLDABLES Write an equation in Point-Slope form and graph the equation under the tab for Lesson 7-2.

EXAMPLES

Write the point-slope form of an equation for each line passing through the given point and having the given slope.

1 $(-2, 7), m = -\frac{1}{3}$

$y - y_1 = m(x - x_1)$ Point-Slope Form

$y -$ ____ $= -\frac{1}{3}(x -$ ____ $)$ Replace x_1, y_1, and m.

$y -$ ____ $= -\frac{1}{3}(x +$ ____ $)$.

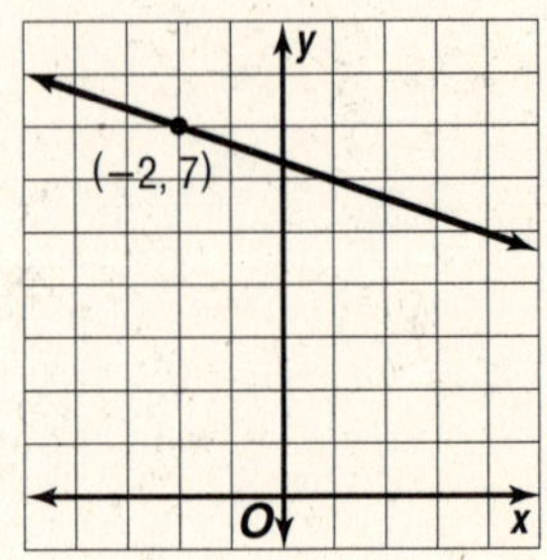

An equation of the line is ____.

2 $(4, 0), m = 4$

$y - y_1 = m(x - x_1)$ Point-Slope Form

$y -$ ____ $= 4(x -$ ____ $)$ Replace x_1, y_1, and m.

____ $= 4(x - 4)$

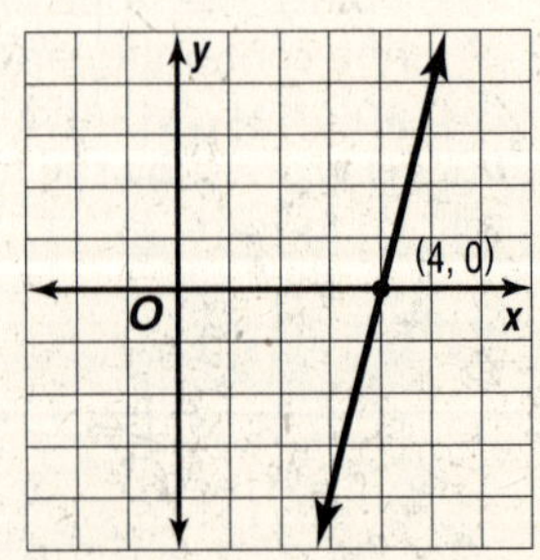

An equation of the line is ____.

Your Turn **Write the point-slope form of an equation for each line passing through the given point and having the given slope.**

a. $(0, -4), m = -\frac{1}{3}$

____

b. $(1, -3), m = 3$

____

Write It

How can you tell that the two methods used in the example provide the same equation?

____

____

____

____

____

____

EXAMPLE

3 Write the point-slope form of an equation of the line shown.

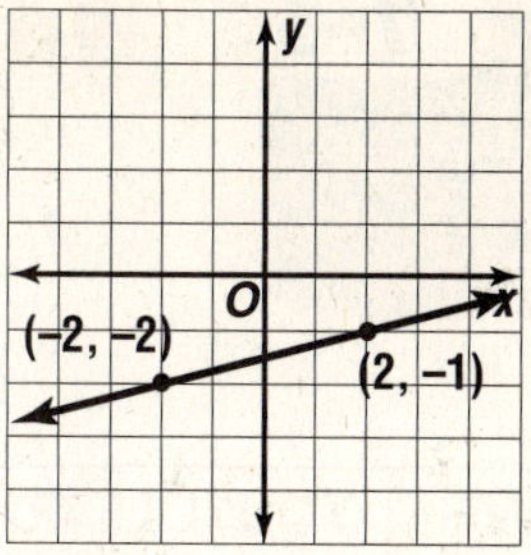

First, determine the slope of the line.

$$m = \frac{y_2 - \underline{\quad}}{\underline{\quad} - x_1} = \frac{\underline{\quad} - (-2)}{2 - \underline{\quad}} \text{ or } \frac{1}{\underline{\quad}}$$

The slope is ____. Use the slope and either point to write an equation.

Method 1 Use (−2, −2).

$y - \underline{\quad} = m(x - \underline{\quad})$ Point-Slope Form

$y - \underline{\quad} = \frac{1}{4}[x - \underline{\quad}]$ $(x_1, y_1) = \underline{\quad}$

$y + \underline{\quad} = \frac{1}{4}(x + \underline{\quad})$

Method 2 Use (2, −1).

$y - \underline{\quad} = m(x - \underline{\quad})$ Point-Slope Form

$y - \underline{\quad} = \frac{1}{4}(x - \underline{\quad})$ $(x_1, y_1) = \underline{\quad}$

$y + \underline{\quad} = \frac{1}{4}(x - \underline{\quad})$

Both ____ and ____ are point-slope forms of an equation for the line passing through (−2, −2) and (2, −1).

Homework Assignment

Page(s): ______________________

Exercises: ______________________

____

Your Turn Write the point-slope form of an equation of the line shown.

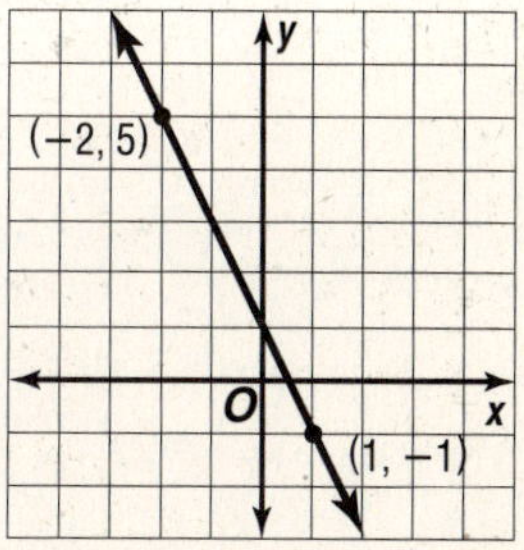

____

7–3 Writing Equations in Slope-Intercept Form

WHAT YOU'LL LEARN

- Write a linear equation in slope-intercept form given the slope and y-intercept.

BUILD YOUR VOCABULARY (page 143)

The y-coordinate of the point at which a graph crosses the ______ is the ***y*-intercept**.

The x-coordinate of the point at which a graph crosses the ______ is the ***x*-intercept**.

KEY CONCEPT

Slope-Intercept Form Given the slope m and y-intercept b of a line, the slope-intercept form of an equation of the line is $y = mx + b$.

FOLDABLES™ Write an equation in slope-intercept form and graph the equation under the tab for Lesson 7-3.

EXAMPLES

Write an equation in slope-intercept form of each line with the given slope and *y*-intercept.

1 $m = 3, b = -1$

$y = mx + b$ Slope-Intercept Form

$= \square x + \square$ Replace m with ______ and b with ______.

An equation of the line is ______.

2 $m = \frac{-2}{3}, b = 0$

$y = mx + b$ Slope-Intercept Form

$= \square x + \square$ Replace m with ______ and b with ______.

An equation of the line is ______.

Your Turn **Write an equation in slope-intercept form of each line with the given slope and *y*-intercept.**

a. $m = 0, b = 5$

b. $m = \frac{1}{2}, b = -3$

EXAMPLES

Write an equation of a line in slope-intercept form for each situation.

3 slope −3 and passes through (1, −4)

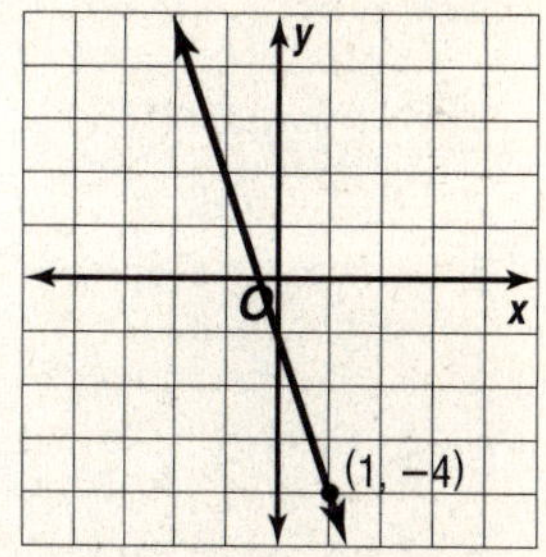

Using the point-slope form, replace (x_1, y_1). Then simplify.

$$y - \square = m(x - \square)$$

$$y - \square = -3(x - \square)$$

$$y + 4 = -3x + 3$$

$$y + 4 - \square = -3x + 3 - \square$$

$$y = -3x - 1$$

An equation of the line is $\square$.

4 passing through (6, 2) and (3, −2)

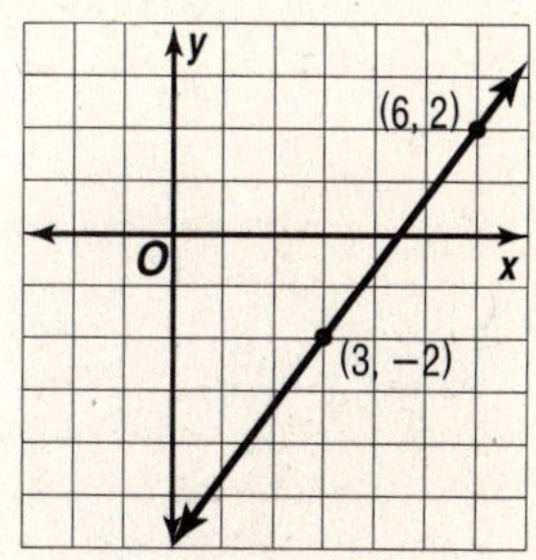

REMEMBER IT

The formula for slope is $m = \frac{y_2 - y_1}{x_2 - x_1}$.

First, determine the slope of the line.

$$m = \frac{-2 - \square}{3 - \square} = \frac{\square}{-3} \text{ or } \frac{4}{3}.$$

Now substitute the known values into the point-slope form.

$y - y_1 = m(x - x_1)$	Point-Slope Form
$y - \square = \square(x - 6)$	Replace (x_1, y_1) with (6, 2) and m with $\frac{4}{3}$.
$y - 2 = \frac{4}{3}x - \square$	Distributive Property

Then write in slope-intercept form.

$y - 2 + 2 = \frac{4}{3}x - 8 + 2$	Add 2 to each side.
$y = \frac{4}{3}x - \square$	Slope-Intercept Form

An equation of the line is $\square$. You can see from the graph that the y-intercept is $\square$.

Your Turn **Write an equation in slope-intercept form of each line.**

a. a line whose slope is $\frac{1}{2}$ and passes through (2, −6)

b. passing through (−1, 4) and (3, 0)

EXAMPLE

5 **A plumber charges $25 for a service call plus $50 per hour of service. The graph represents the plumber's charges. Write an equation of the line in slope-intercept form.**

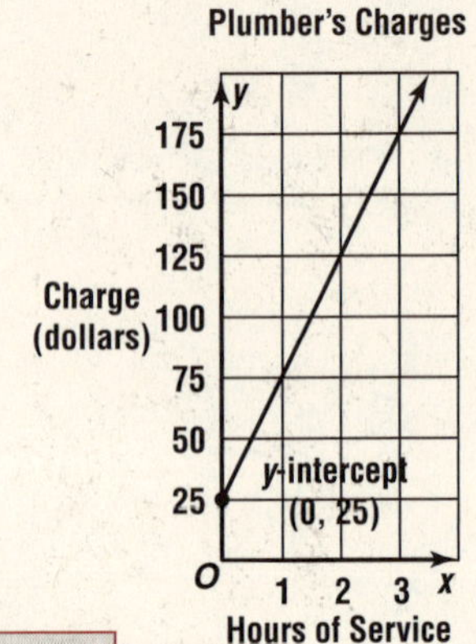

The *y*-intercept of the line is 25.

Determine the slope.

$$m = \frac{y_2 - \square}{\square - x_1} = \frac{75 - \square}{1 - \square} \text{ or } \square$$

Now substitute these values into the ______-intercept form.

$y = mx + b$	Slope-Intercept Form
$= \square x + \square$	Replace *m* with ______ and *b* with ______.

An equation of the line is ______.

Your Turn A taxi driver charges $5 for each ride plus $2 per mile. The graph represents the taxi driver's charges. Write an equation of the line in slope intercept form.

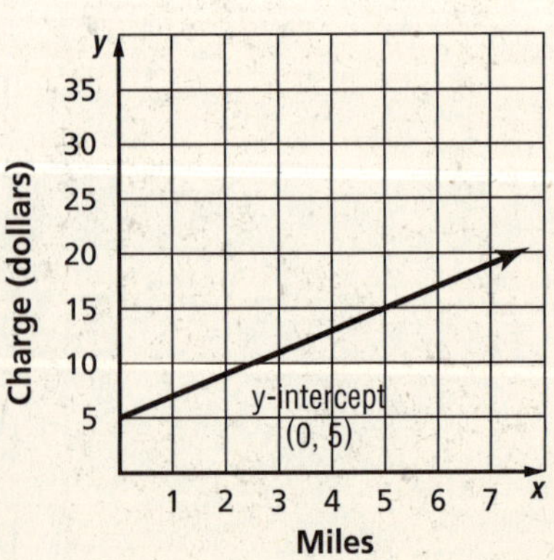

HOMEWORK ASSIGNMENT

Page(s):

Exercises:

7–4 Scatter Plots

What You'll Learn

- Graph and interpret points on scatter plots.

Build Your Vocabulary (page 143)

A graph in which data is displayed as ordered pairs is a **scatter plot**.

Key Concepts

Scatter Plots

Positive relationship: As *x* increases, *y* increases.

Negative relationship: As *x* increases, *y* decreases.

No relationship: No obvious pattern.

EXAMPLE

1 The scatter plot shows the number of years of experience and the salary for each employee in a small company. Determine whether the scatter plot shows a positive relationship, negative relationship, or no relationship. If there is a relationship, describe it.

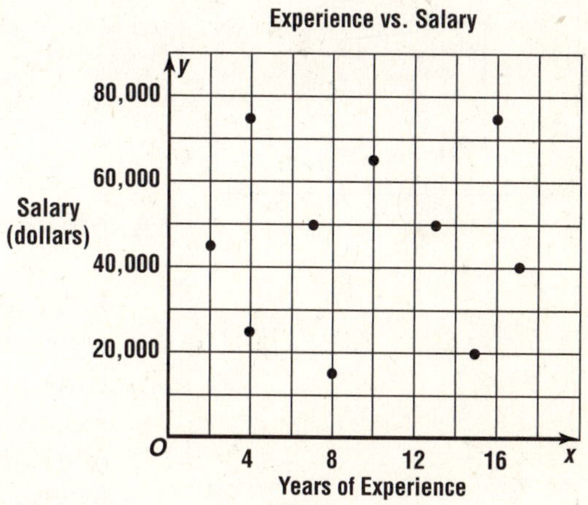

As the number of years of experience increase, the salary

does not seem to increase or ________. Thus, there is

________ between experience and ________.

Your Turn Use the scatter plot shown. Determine whether the scatter plot shows a positive relationship, negative relationship, or no relationship. If there is a relationship, describe it.

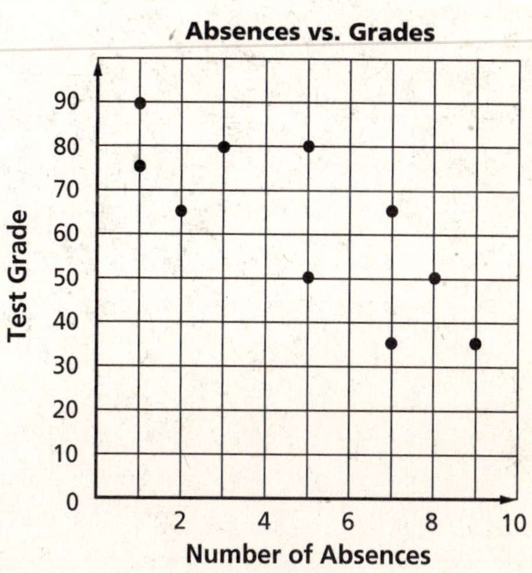

EXAMPLE

FOLDABLES™

ORGANIZE IT

Give an example of a scatter plot that has a positive relationship under the tab for Lesson 7-4. Also give an example of one that has a negative relationship and one that has no relationship.

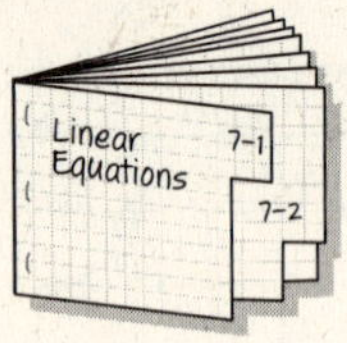

2 **The table shows the average number of minutes a pediatric dentist spends during each appointment instructing the patient in proper dental care, and the number of cavities for each patient.**

Instruction Time (min)	Number of Cavities
6	1
4	3
7	2
10	1
1	5
1	6
5	3
2	4
2	3

A. Make a scatter plot of the data.

Let the horizontal axis represent the instruction time and let the vertical axis represent the number of cavities. Plot the data as shown.

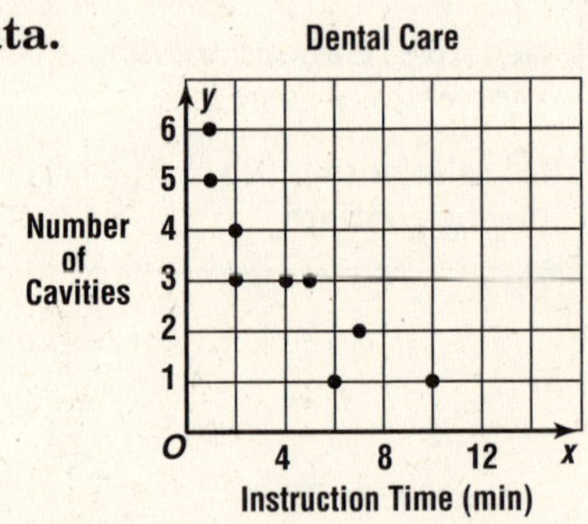

B. Does the scatter plot show a relationship between instruction time and cavities? Explain.

_____ ; it appears that a _____ number of cavities is directly related to a _____ amount of instruction time. There is a _____ relationship.

C. Describe the independent and dependent variables. Then state the domain and the range.

The number of _____ depends on the _____ , so the number of cavities is the _____ variable. The _____ is the set of all instruction times and the _____ is the set of all numbers of cavities.

Your Turn **The table shows the gestation period and average longevity for various animals.**

Animal	Gestation (days)	Longevity (years)
kangaroo	36	7
leopard	98	12
tiger	105	16
baboon	187	20
hippopotamus	238	41
gorilla	258	20

A. Make a scatter plot of the data. Let the horizontal axis represent gestation time and let the vertical axis represent the longevity of the animal.

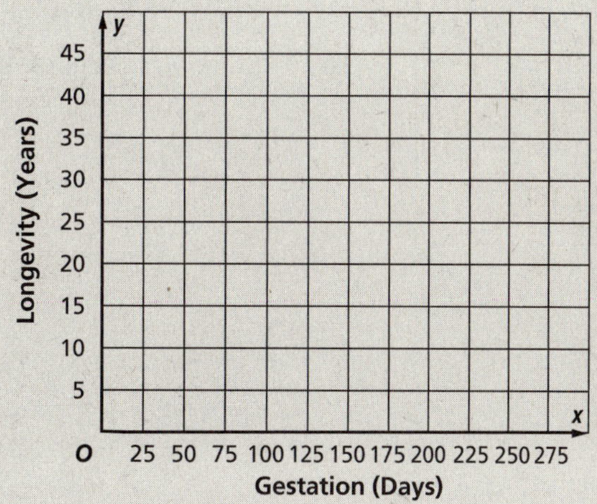

B. Does the scatter plot show a relationship between gestation period and longevity? Explain.

C. Describe the independent and dependent variables. Then state the domain and the range.

Remember It

The domain is the set of all of the first coordinates in a set of ordered pairs and the range is the set of all of the second coordinates in the set of ordered pairs.

Homework Assignment

Page(s):

Exercises:

7–5 Graphing Linear Equations

WHAT YOU'LL LEARN

- Graph linear equations by using the *x*- and *y*- intercepts or the slope and *y*-intercept.

FOLDABLES™

ORGANIZE IT

Draw the graph of a line and label the *x* and *y* intercepts under the tab for Lesson 7-5.

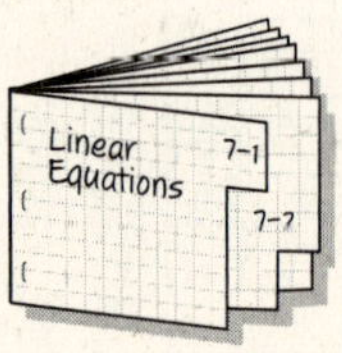

EXAMPLE

1 Determine the *x*-intercept and *y*-intercept of the graph of $2y - x = 8$. Then graph the equation.

To find the *x*-intercept, let $y = 0$.

$2y - x = 8$

$2(0) - x = 8$ Replace *y* with 0.

____ $= 8$

$\frac{-x}{-1} = \frac{8}{-1}$ Divide each side by −1.

$x =$ ____

To find the *y*-intercept, let $x = 0$.

$2y - x = 8$

$2y - 0 = 8$ Replace *x* with 0.

____ $= 8$

$\frac{2y}{2} = \frac{8}{2}$ Divide each side by 2.

$y =$ ____

The *x*-intercept is ____, and the *y*-intercept is ____. This means that the graph intersects the *x*-axis at ____ and the *y*-axis at ____. Graph the equation.

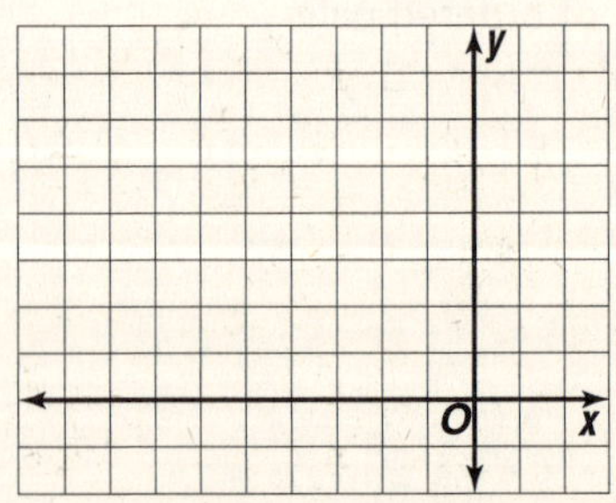

Your Turn Determine the *x*-intercept and *y*-intercept of the graph. Graph the equation. $3y + 2x = -6$

____

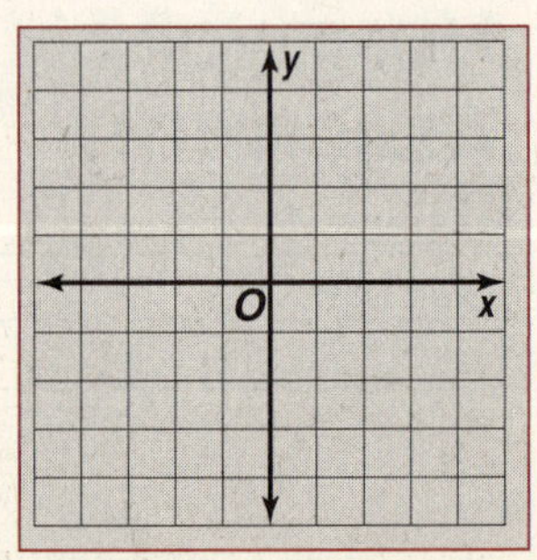

EXAMPLES

Graph each equation by using the slope and y-intercept.

REMEMBER IT

A negative slope means that the graph of the line decreases when tracing the line from left to right. A positive slope means that the graph of the line increases when tracing the line from left to right.

2 $y = -\frac{1}{5}x + 2$

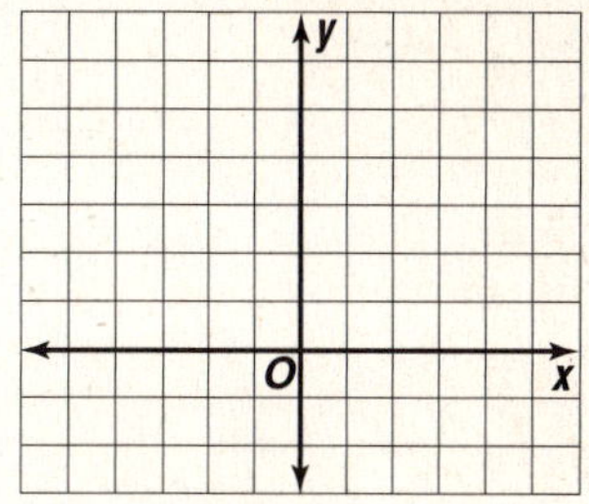

$$y = mx + b$$

$$y = -\frac{1}{5}x + 2$$

The slope is ______, and the y-intercept is ______. Graph the point at ______. Then go ______ 1 unit and ______ 5 units. This will be the point at ______. Then draw the line through points ______ and ______.

3 $3x - y = 4$

First, write the equation in slope-intercept form.

$$3x - y = 4$$

$$3x - y - 3x = 4 - 3x \qquad \text{Subtract } 3x \text{ from each side.}$$

$$-y = \text{\_\_\_\_\_\_} + 4$$

$$\frac{-y}{\text{\_\_\_\_\_\_}} = \frac{-3x + 4}{\text{\_\_\_\_\_\_}} \qquad \text{Divide each side by \_\_\_\_\_\_.}$$

$$y = \text{\_\_\_\_\_\_} - 4$$

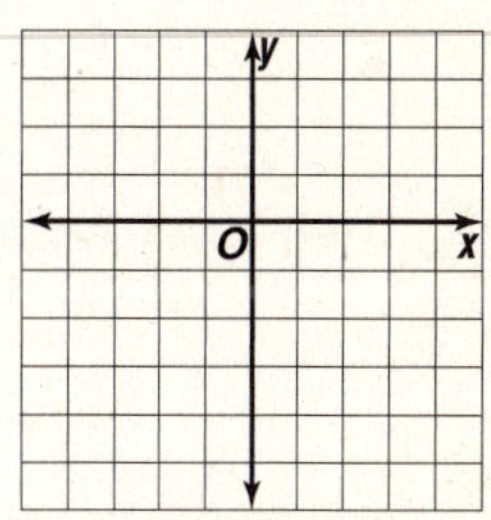

The slope is ______. The y-intercept is ______. Graph the point at ______. Then go ______ 3 units and ______ 1 unit. This will be the point at ______. Then draw the line through points ______ and ______.

Your Turn **Graph each equation by using the slope and *y*-intercept.**

a. $y = \frac{1}{3}x - 1$

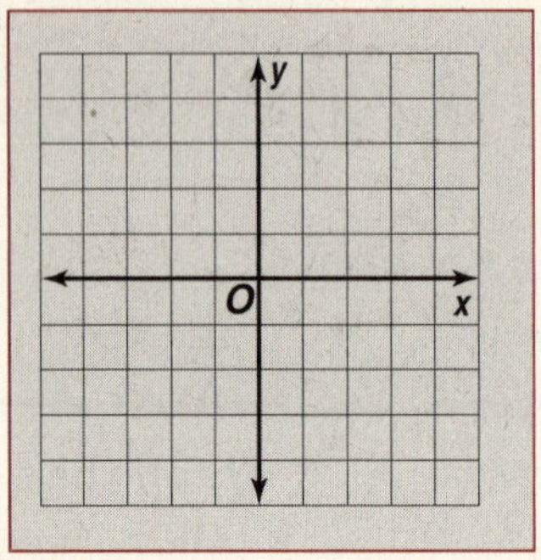

b. $2x + 4y = 8$

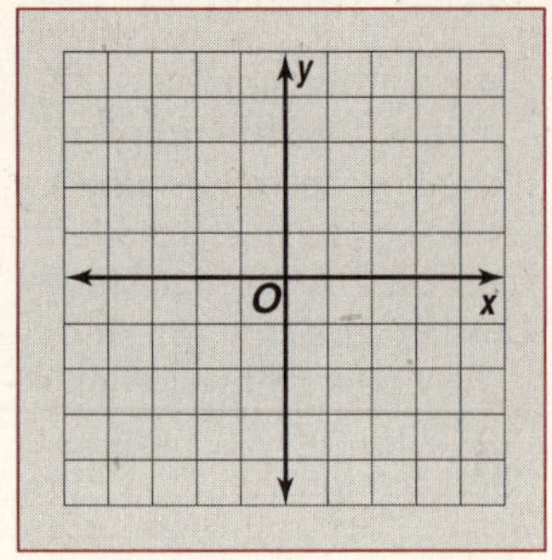

EXAMPLE

4 Graph $y = -3$.

$y = mx + b$ Slope-Intercept Form

$y = 0x - 3$ slope = ____, *y*-intercept = ____

No matter what the value of x,

$y =$ ____. So, all ordered pairs

are of the form (x, ____). Some

examples are (0, ____) and

(2, ____).

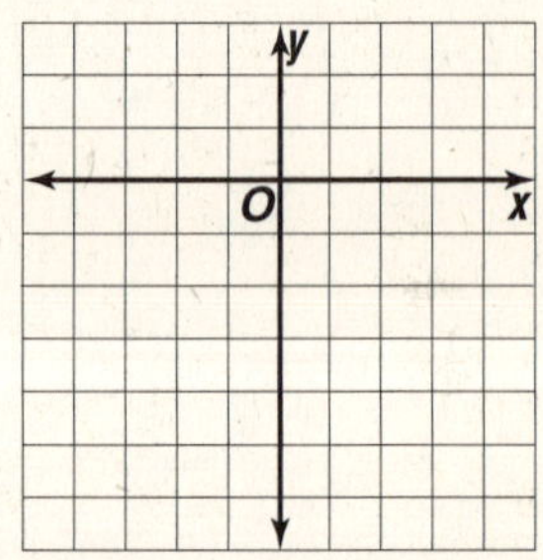

Your Turn **Graph each equation.**

a. $x = -4$

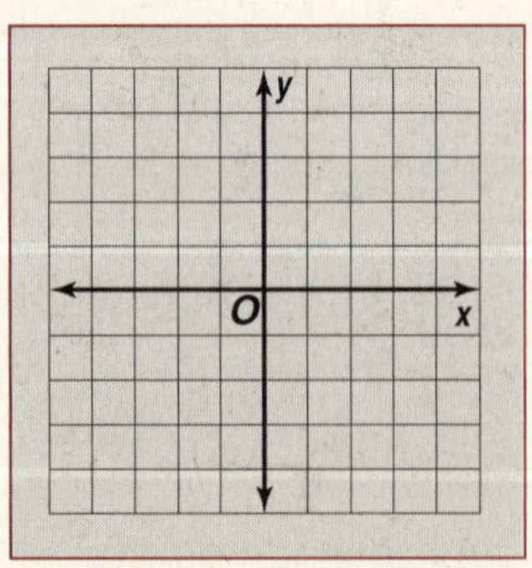

b. $y = 2$

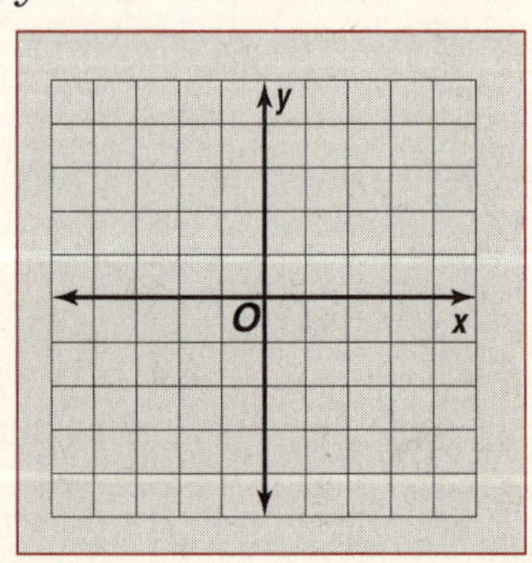

HOMEWORK ASSIGNMENT

Page(s): ____

Exercises: ____

7–6 Families of Linear Graphs

What You'll Learn

- Explore the effects of changing the slopes and y-intercepts of linear functions.

Build Your Vocabulary (page 142)

Graphs and equations of graphs that have at least one characteristic in common are a **family of graphs**.

Foldables™

Organize It

Summarize the two categories of families of graphs under the tab for Lesson 7-6. What is the effect of a change in the slope and y-intercept on the graph of a line?

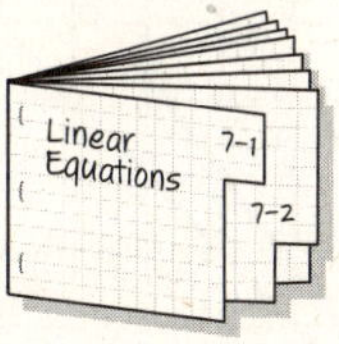

Examples

Graph each pair of equations. Describe any similarities or differences. Explain why they are a family of graphs.

1 $y = -\frac{1}{2}x + 2$

$y = -\frac{1}{2}x - 1$

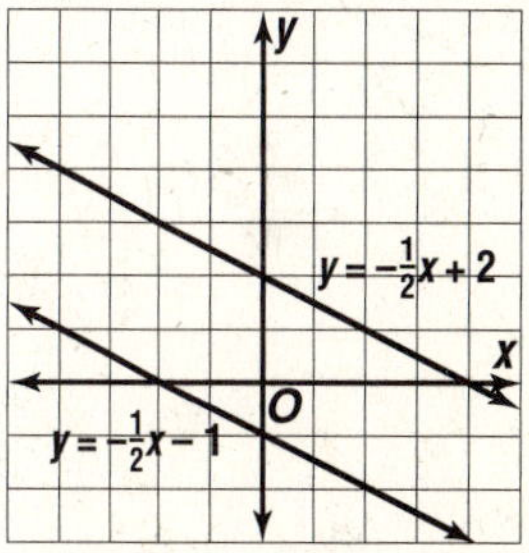

The graphs have y-intercepts of 2, and -1. They are a family of graphs because each slope is ______.

2 $y = 5x - 1$

$y = -x - 1$

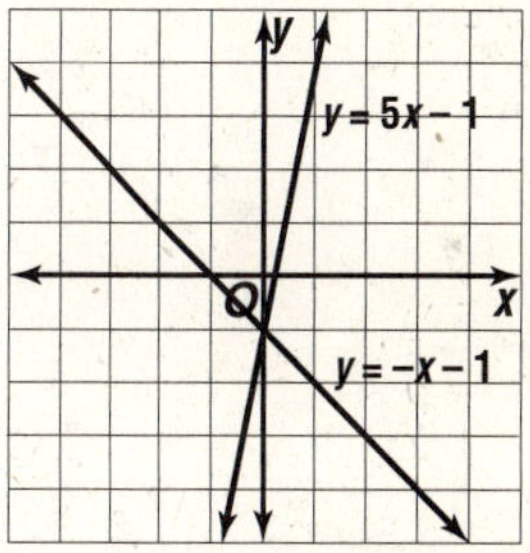

Each graph has a ______ slope.

Each graph has a y-intercept of -1. Thus, they are a family of graphs.

Your Turn **Graph each pair of equations. Describe any similarities or differences. Explain why they are a family of graphs.**

a. $y = x - 3$

$y = x + 2$

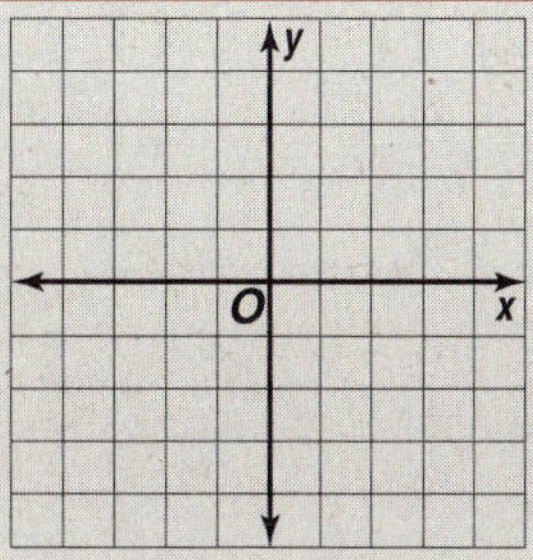

b. $y = 3x - 2$

$y = \frac{2}{3}x - 2$

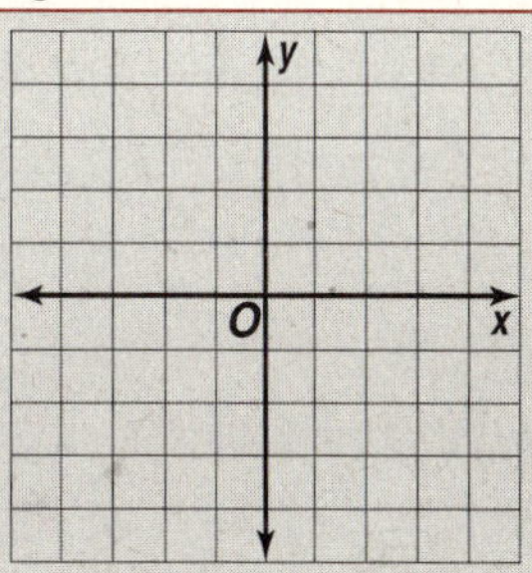

EXAMPLE

3 Gretchen and Max each have a savings account and plan to save $20 per month. The current balance in Gretchen's account is $150 and the balance in Max's account is $100. Then $y = 20x + 150$ and $y = 20x + 100$ represent how much money each has in their account, respectively, after x months. Compare and contrast the graphs of the equations.

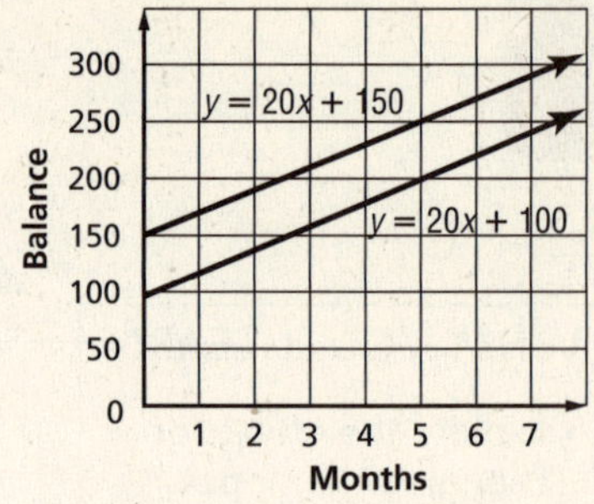

The equations have the same ______, but the ______ of Gretchen's graph is ______ than the y-intercept of Max's graph. Gretchen's account will always have more money than Max's.

Your Turn Tyler and Ying both babysit for Mrs. Hernandez. Tyler charges $4 per hour and Ying charges $5 per hour. Suppose x represents the number of hours. Then $y = 4x$ and $y = 5x$ represent how much money each person will make, respectively, after x hours. Compare and contrast the graphs of the equations.

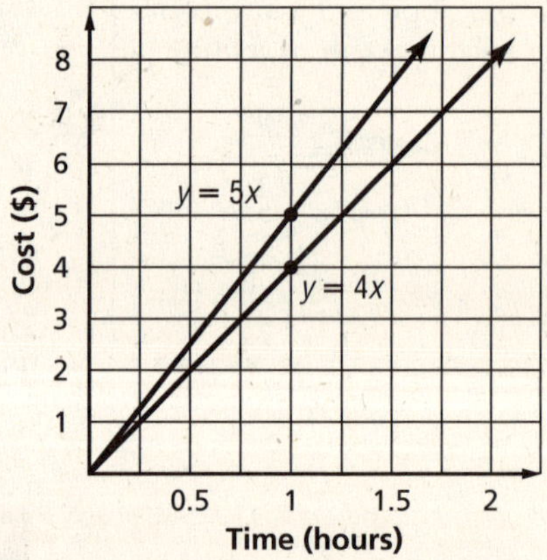

BUILD YOUR VOCABULARY (page 142)

The simplest of graphs in a ______ of graphs is a **parent** graph.

EXAMPLES

Change $y = -3x - 1$ so that the graph of the new equation fits each description.

4 same y-intercept, less steep positive slope

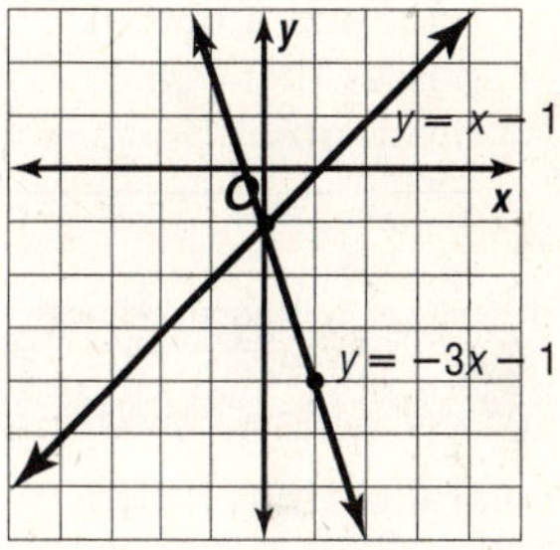

The y-intercept is ______, and the slope slope is ______. The new equation will also have a ______ of −1. In order for the slope to be less steep and positive, its value must be ______ than 3, such as 1. The new equation is ______.

5 same slope, y-intercept is shifted down 2 units

The slope of the new equation will be ______. Since the current y-intercept is ______, the new y-intercept will be $-1 - 2$, or ______. The new equation is ______. Check by graphing.

KEY CONCEPT

Parent Graphs

***y* = *x*:** As the value of *m* increases, the line gets steeper.

***y* = −*x*:** As the value of *m* decreases, the line gets steeper.

***y* = 2*x*:** As the value of *b* increases, the graph shifts up on the *y*-axis. As the value of *b* decreases, the graph shifts down on the *y*-axis.

Your Turn **Change $y = -x + 3$ so that the graph of the new equation fits each description.**

a. same slope, y-intercept is shifted up 4 units

______

b. same y-intercept, steeper negative slope

______

HOMEWORK ASSIGNMENT

Page(s):

Exercises:

7–7 Parallel and Perpendicular Lines

EXAMPLE

WHAT YOU'LL LEARN

- Write an equation of a line that is parallel or perpendicular to the graph of a given equation and that passes through a given point.

KEY CONCEPT

Parallel Lines If two lines have the same slope, then they are parallel.

FOLDABLES Give two examples of equations whose graphs are parallel and two examples of equations whose graphs are perpendicular under the tab for Lesson 7-7.

1 Determine whether the graphs of the equations are parallel.

$\mathbf{y = 3x + 4}$

$\mathbf{9x + 3y = 12}$

First, determine the slope of each line. Write each equation in slope-intercept form.

$y = 3x + 4$ Slope-Intercept Form

The slope is ______.

$9x + 3y = 12$

$9x + 3y -$ ______ $= 12 -$ ______

$\frac{3y}{\text{\_\_\_\_}} = \frac{-9x + 12}{\text{\_\_\_\_}}$ Divide each side by ______.

$y =$ ______ $x +$ ______

The slope is ______.

The slopes are not the same so the lines are not ______.

Your Turn Determine whether the graphs of the equations are parallel.

$3x + y = 2 \qquad 2y = -6x - 8$

EXAMPLE

2 **Write an equation in slope-intercept form of the line that is parallel to the graph of $y = \frac{2}{3}x - 3$ and passes through the point at (−3, 1).**

The slope of the given line is ______. So, the slope of the new line will also be ______. Find the new equation by using the point-slope form.

$y - y_1 = m(x - x_1)$	Point-Slope Form
$y -$ ______ $= \frac{2}{3}(x -$ ______ $)$	$x_1 =$ ______, $y_1 =$ ______, and $m =$ ______.
$y - 1 = \frac{2}{3}x +$ ______	Distributive Property
$y - 1 + 1 = \frac{2}{3}x + 2 + 1$	Add 1 to each side.
$y = \frac{2}{3}x +$ ______	

An equation whose graph is parallel to the graph of $y = \frac{2}{3}x - 3$ and passes through (−3, 1) is ______.

Your Turn Write an equation in slope-intercept form of the line that is parallel to the graph of $y = \frac{1}{2}x - 4$ and passes through the point at (−6, 2).

KEY CONCEPT

Perpendicular Lines
If the product of the slopes of two lines is −1, then the lines are perpendicular.

EXAMPLE

3 **Determine whether the graphs of the equations are perpendicular.**

$\mathbf{y + 2x = 4}$

$\mathbf{y = \frac{1}{2}x + 3}$

First, determine the slopes of the lines. Write each equation in slope-intercept form.

$$y + 2x = 4 \qquad\qquad y = \frac{1}{2}x + 3$$

$$y + 2x - \square = 4 - \square$$

$$y = \square + 4$$

The graphs are perpendicular because the product of their slopes is $-2 \cdot \frac{1}{2}$ or $\square$.

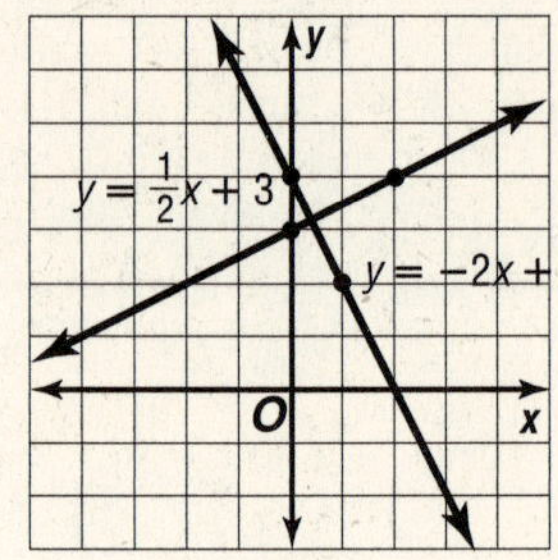

Your Turn Determine whether the graphs of $x + y = 1$ and $x - y = -4$ are perpendicular.

EXAMPLE

4 **Write an equation in slope-intercept form of the line that is perpendicular to the graph of $y = -2x + 5$ and passes through the point at (2, −3).**

The slope is −2. A line perpendicular to the graph of $y = -2x + 5$ has slope $\frac{1}{2}$. Find the new equation by using the point-slope form.

$y - y_1 = m(x - x_1)$	Point-Slope Form
$y - \square = \frac{1}{2}(x - \square)$	$x_1 = \square$, $y_1 = \square$, and $m = \square$.
$y + \square = \frac{1}{2}x - \square$	Distributive Property
$y + 3 - 3 = \frac{1}{2}x - 1 - 3$	Subtract 3 from each side.
$y = \frac{1}{2}x - \square$	

Your Turn Write an equation in slope-intercept form of the line that is perpendicular to the graph of $y = \frac{1}{4}x + 2$ and passes through the point at (−1, −2).

HOMEWORK ASSIGNMENT

Page(s):

Exercises:

CHAPTER 7 BRINGING IT ALL TOGETHER

STUDY GUIDE

FOLDABLES™	VOCABULARY PUZZLEMAKER	BUILD YOUR VOCABULARY
Use your **Chapter 7 Foldable** to help you study for your chapter test.	To make a crossword puzzle, word search, or jumble puzzle of the vocabulary words in Chapter 7, go to: www.glencoe.com/sec/math/t_resources/free/index.php	You can use your completed **Vocabulary Builder** (pages 142–143) to help you solve the puzzle.

7-1 Slope

Find the slope between each set of points.

1. (2, −3), (5, 1) **2.** (0, 5), (−1, −4) **3.** (2, 0), (2, −1)

7-2 Writing Equations in Point-Slope Form

4. In the formula $y - y_1 = m(x - x_1)$, what do x_1 and y_1 represent?

Match each equation with the correct graph.

5. $y - 3 = \frac{1}{2}(x + 2)$ **6.** $y = 2x$ **7.** $y + 1 = -\frac{2}{3}x$

a.

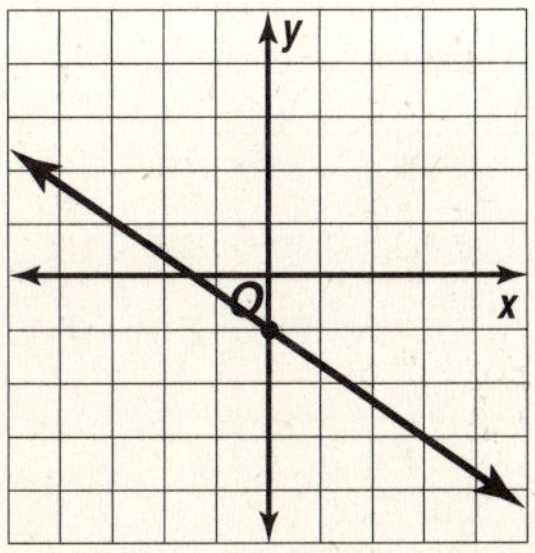

b.

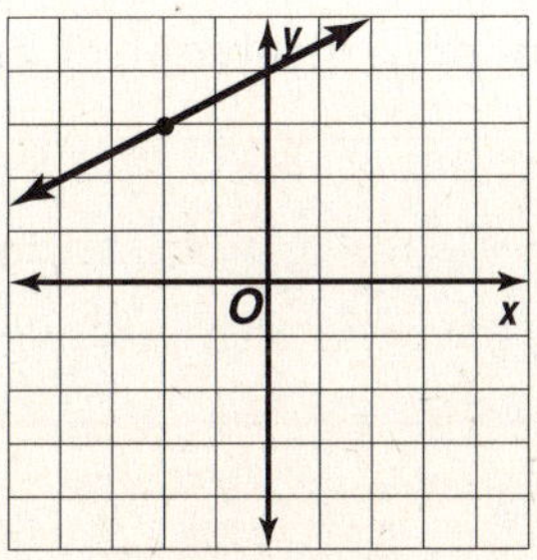

c.

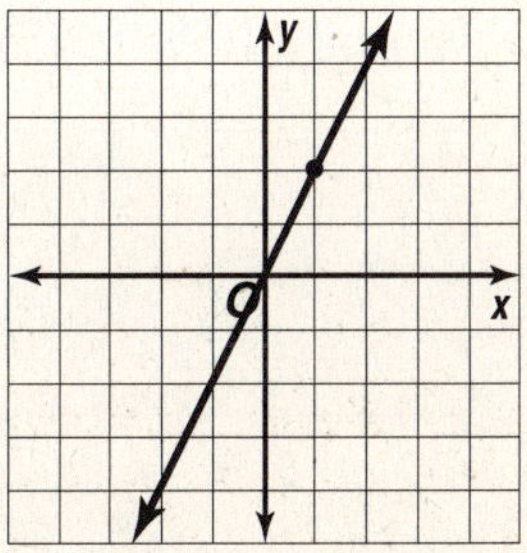

7-3

Writing Equations in Slope-Intercept Form

Complete the chart below by writing the formula for each form of equation. Then write an example of each equation.

	Form of Equation	Formula	Example
8.	standard form		
9.	slope-intercept form		

7-4

Scatter Plots

Refer to the scatter plot shown at the right.

10. Which quantity is the independent quantity? the dependent quantity?

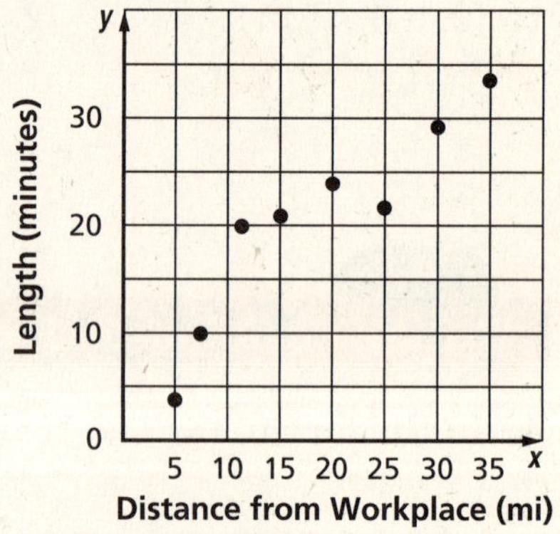

11. What conclusion can you draw from the scatter plot?

7-5

Graphing Linear Equations

12. Explain how to find the x- or y-intercept of an equation.

Refer to the graph shown at the right.

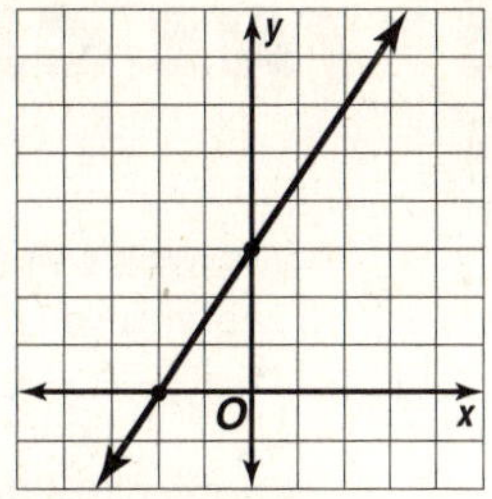

13. What is the y-intercept of the graph?

14. What is the x-intercept of the graph?

7-6 Families of Linear Graphs

Tell whether each set of graphs is a family of graphs.

15.

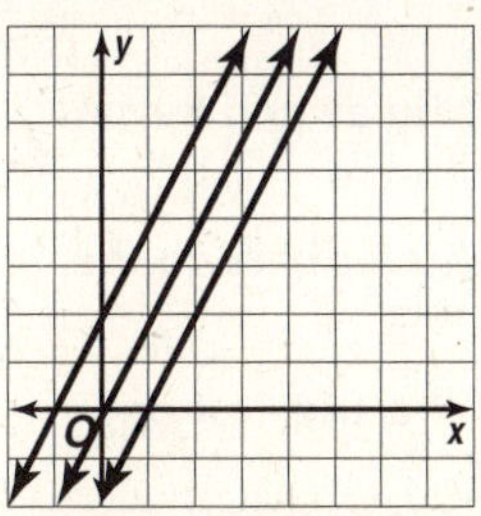

16.

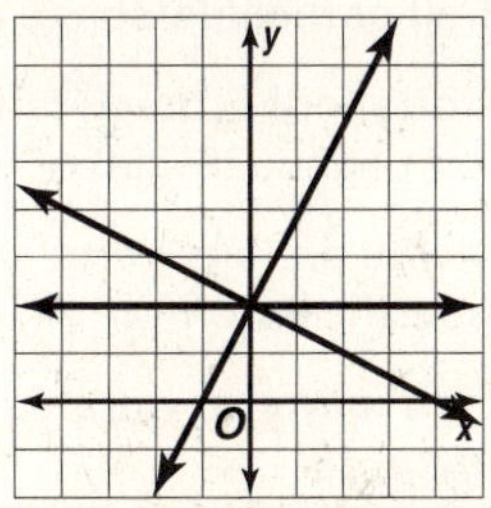

7-7 Parallel and Perpendicular Lines

17. Parallel lines ______ (always/never) intersect.

Tell whether the graphs are *parallel, perpendicular,* or *neither.* Explain.

18.

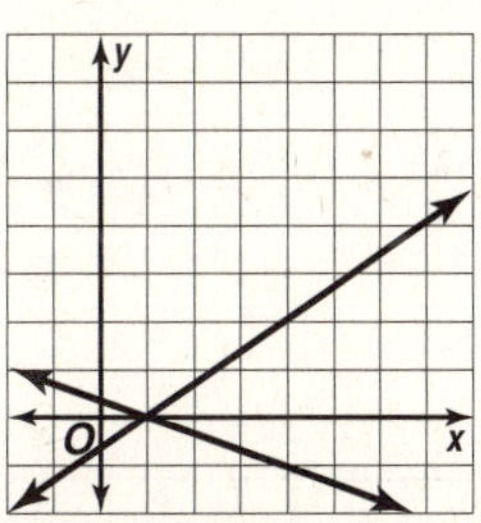

19.

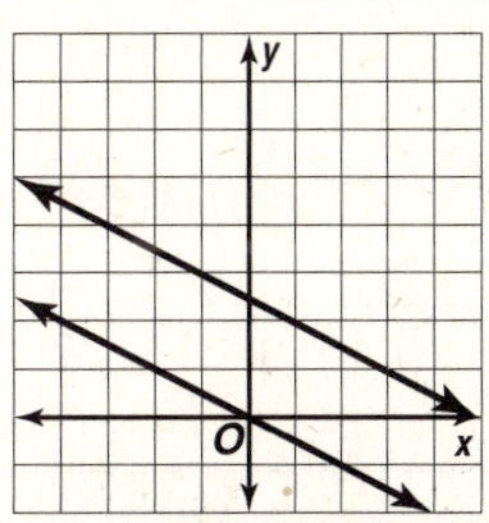

CHAPTER 7

Checklist

ARE YOU READY FOR THE CHAPTER TEST?

Math Online

Visit **algconcepts.com** to access your textbook, more examples, self-check quizzes, and practice tests to help you study the concepts in Chapter 7.

Check the one that applies. Suggestions to help you study are given with each item.

☐ **I completed the review of all or most lessons without using my notes or asking for help.**

- You are probably ready for the Chapter Test.
- You may want take the Chapter 7 Practice Test on page 331 of your textbook as a final check.

☐ **I used my Foldable or Study Notebook to complete the review of all or most lessons.**

- You should complete the Chapter 7 Study Guide and Review on pages 328–330 of your textbook.
- If you are unsure of any concepts or skills, refer back to the specific lesson(s).
- You may also want to take the Chapter 7 Practice Test on page 331.

☐ **I asked for help from someone else to complete the review of all or most lessons.**

- You should review the examples and concepts in your Study Notebook and Chapter 7 Foldable.
- Then complete the Chapter 7 Study Guide and Review on pages 328–330 of your textbook.
- If you are unsure of any concepts or skills, refer back to the specific lesson(s).
- You may also want to take the Chapter 7 Practice Test on page 331.

Student Signature

Parent/Guardian Signature

Teacher Signature

CHAPTER 8

Powers and Roots

FOLDABLES™ Use the instructions below to make a Foldable to help you organize your notes as you study the chapter. You will see Foldable reminders in the margin of this Interactive Study Notebook to help you in taking notes.

Begin with four sheets of grid paper.

STEP 1 **Fold**
Fold each sheet of grid paper in half along the width. Then cut along the crease.

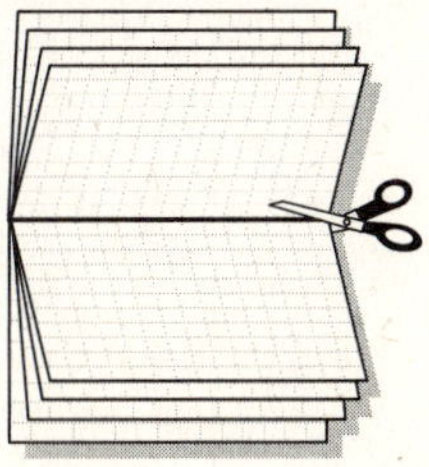

STEP 2 **Staple**
Staple the eight half-sheets together to form a booklet.

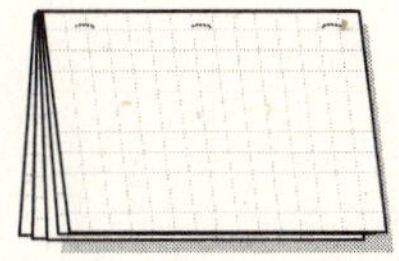

STEP 3 **Cut**
Cut even lines from the bottom of the top sheet, six lines from the second sheet, and so on.

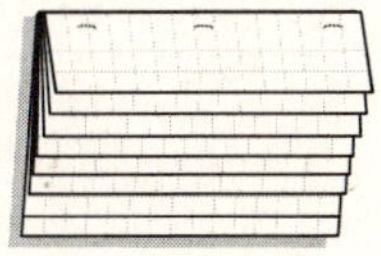

STEP 4 **Label**
Label the tabs with lesson topics as shown.

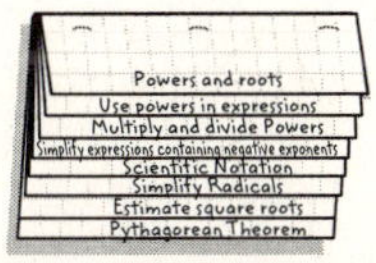

NOTE-TAKING TIP: When taking notes in math class, be sure to write down important rules and properties. It is also a good idea to record examples of any rules and properties.

Chapter 8

CHAPTER 8

BUILD YOUR VOCABULARY

This is an alphabetical list of new vocabulary terms you will learn in Chapter 8. As you complete the study notes for the chapter, you will see Build Your Vocabulary reminders to complete each term's definition or description on these pages. Remember to add the textbook page number in the second column for reference when you study.

Vocabulary Term	Found on Page	Definition	Description or Example
base			
composite number [kahm-PA-zit]			
converse			
exponent [ek-SPO-nent]			
hypotenuse [hi-PA-tin-oos]			
irrational numbers [i-RA-shun-ul]			
leg			
negative exponent			

Vocabulary Term	Found on Page	Definition	Description or Example
perfect square			
power			
prime factorization [FAK-tor-i-ZAY-shun]			
prime number			
Pythagorean Theorem [puh-THA-guh-REE-un]			
radical [RAD-ik-ul]			
radical expression			
radical sign			
scientific notation			
square root			

8–1 Powers and Exponents

WHAT YOU'LL LEARN

- Use powers in expressions.

BUILD YOUR VOCABULARY (pages 168–169)

A **perfect square** is the ______ of an integer and itself.

An **exponent** tells how many times a number is used as a ______.

The number that is raised to a ______ is called a **base**.

A **power** is a number that is expressed when using an ______.

EXAMPLES

Write each expression using exponents.

1 $5 \cdot 5 \cdot 5$

The base is ______. It is a factor ______ times. $5 \cdot 5 \cdot 5 =$ ______

2 $d \cdot d \cdot d \cdot d \cdot d \cdot d$

The base is ______. It is a factor ______ times.

$d \cdot d \cdot d \cdot d \cdot d \cdot d =$ ______

EXAMPLE

3 **Write (6)(6)(−7)(−7)(−7)(−7)(−7) using exponents.**

Use the Associative Property to group factors with like bases.

$(6)(6)(-7)(-7)(-7)(-7)(-7) = [(6)(6)][(-7)(-7)(-7)(-7)(-7)]$

$= $ ______

Your Turn **Write each expression using exponents.**

a. c

______

b. $(-3)(-3)(-3)(-3)$

______

FOLDABLES™

ORGANIZE IT

Write three examples of expressions with exponents under the appropriate tab. Then, write them again in expanded form.

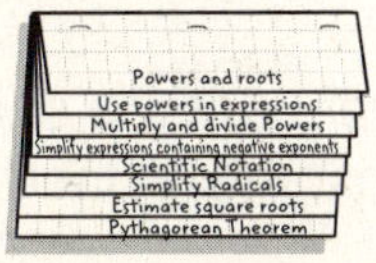

KEY CONCEPT

Order of Operations

1. Do all operations within grouping symbols first; start with the innermost grouping symbols.
2. Evaluate all powers in order from left to right.
3. Do all multiplications and divisions from left to right.
4. Do all additions and subtractions from left to right.

HOMEWORK ASSIGNMENT

Page(s):

Exercises:

EXAMPLES

Write each power as a multiplication expression.

4 6^4

$6^4 =$ ______ The exponent 4 means that 6 is a factor 4 times.

5 h^5

$h^5 =$ ______ The exponent 5 means that h is a factor 5 times.

6 $7a^3b^2$

$7a^3b^2 =$ ______ 7 is used as a factor once, a is used 3 times, and b is used twice.

Your Turn **Write each power as a multiplication expression.**

a. 8^3

b. b^5

c. $3yz^4$

EXAMPLE

7 **Evaluate $5a^2$ if $a = 4$.**

$5a^2 = 5(\square)^2$ Replace a with 4.

$= 5(\square)$ Evaluate the power $4 \cdot 4 = 16$.

$= \square$ Multiply.

Your Turn **Evaluate each expression.**

a. $2x^3$ if $x = 4$

b. $4b^2 - c^3$ if $b = -2$ and $c = -2$

8–2 Multiplying and Dividing Powers

EXAMPLES

What You'll Learn

- Multiply and divide powers.

Key Concepts

Product of Powers You can multiply powers with the same base by adding the exponents.

Quotient of Powers You can divide powers with the same base by subtracting the exponents.

Simplify each expression.

1 $5^2 \cdot 5^7$

$5^2 \cdot 5^7 = 5^{\square}$

$= \square$

To multiply powers that have the same base, write the common base, then add the exponents.

2 $s^{10} \cdot s^5$

$s^{10} \cdot s^5 = s^{\square}$

$= \square$

To multiply powers that have the same base, write the common base, then add the exponents.

3 $(10a)(5a^4)$

$(10a)(5a^4) = (\square \cdot 5)(\square \cdot a^4)$

$= 50a^{\square}$

$= \square$

Use the Commutative and Associative Properties. $a = a^1$

4 $(m^5n^4)(m^3n^7)$

$(m^5n^4)(m^3n^7) = (m^5 \cdot m^3)(n^4 \cdot n^7)$

$= m^{\square} \cdot n^{\square}$

$= \square$

Use the Commutative and Associative Properties.

Your Turn **Simplify each expression.**

a. $5^3 \cdot 5^6$

b. $a \cdot a^4$

c. $(6x)(3x^2)$

d. $(a^2b^3)(a^8b^5)$

EXAMPLES

FOLDABLES™

ORGANIZE IT

Write three examples of multiplying terms with powers and three examples of dividing terms with powers under the tab for Lesson 8-2.

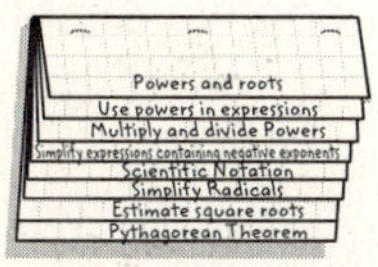

Simplify each expression.

5 $\dfrac{7^6}{7^2}$

$\dfrac{7^6}{7^2} = 7^{\square - 2}$

$= \square$

To divide powers that have the same base, write the common base. Then subtract the exponents

6 $\dfrac{p^{10}}{p}$

$\dfrac{p^{10}}{p} = p^{\square - 1}$

$= \square$

Write the common base. Then subtract the exponents.

7 $\dfrac{15a^6b^4}{3a^4b^3}$

$\dfrac{15a^6b^4}{3a^4b^3} = \left(\dfrac{\square}{3}\right)\left(\dfrac{a^6}{a^4}\right)\left(\dfrac{b^4}{\square}\right)$

Group the powers that have the same base.

$= 5a^{\square}b^{\square}$

$= 5a^{\square}b^{\square}$

$= \square$

Your Turn **Simplify each expression.**

a. $\dfrac{10^8}{10^5}$

b. $\dfrac{c^7}{c}$

c. $\dfrac{20x^5y^6}{5x^2y}$

d. $\dfrac{21y^8z^9}{3y^2z^8}$

HOMEWORK ASSIGNMENT

Page(s):

Exercises:

8–3 Negative Exponents

WHAT YOU'LL LEARN

- Simplify expressions containing negative exponents.

KEY CONCEPT

Negative Exponents

$a^{-n} = \frac{1}{a^n}$; $\frac{1}{a^{-n}} = a^n$

The value of *a* cannot be 0.

FOLDABLES™

ORGANIZE IT

Write three examples of variables with negative exponents under the tab for Lesson 8-3. Then, write those examples using positive exponents.

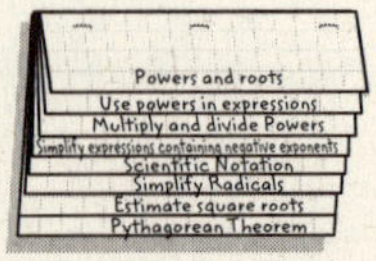

EXAMPLE

1 **Write 6^{-2} using positive exponents. Then evaluate the expression.**

$6^{-2} = \frac{1}{6^{\square}}$ Definition of negative exponent.

$= \square$ $6 \cdot 6 = \square$

EXAMPLES

Simplify each expression.

2 q^3r^{-4}

$q^3r^{-4} = q^3 \cdot r^{-4}$

$= q^3 \cdot \frac{1}{\square}$ Definition of negative exponent.

$= \square$

3 $\frac{m^2n^{10}}{m^5n^2}$

$\frac{m^2n^{10}}{m^5n^2} = \left(\frac{m^2}{m^5}\right) \cdot \left(\frac{n^{10}}{n^2}\right)$

$= m^{\square} \cdot n^{\square}$ Quotient of powers.

$= m^{\square} \cdot n^{\square}$

$= \frac{1}{\square} \cdot n^8$ Definition of negative exponent.

$= \square$

Your Turn **Simplify each expression.**

a. x^{-3} $\square$ **b.** $a^{-3}b$ $\square$ **c.** $\frac{c^2d^5}{c^3d}$ $\square$

EXAMPLE

4 **Simplify $\frac{-10h^{-6}k^4}{25h^3k^7}$.**

$$\frac{-10h^{-6}k^4}{25h^3k^7} = \left(\frac{-10}{25}\right)\left(\frac{h^{-6}}{h^3}\right)\left(\frac{k^4}{k^7}\right)$$

$$= \left(\frac{-2}{\square}\right)\left(\frac{h^{-6}}{h^3}\right)\left(\frac{k^4}{k^7}\right)$$ Write $\frac{-10}{25}$ in simplest form.

$$= \frac{-2h^{\square}k^{\square}}{5}$$ Quotient of powers.

$$= \frac{-2h^{\square}k^{\square}}{5}$$ Simplify the powers.

$$= \square$$

EXAMPLE

5 **A large archery target has a diameter of 1 meter. An arrow tip has a diameter of 10^{-2} meter. How many arrows could fit across the diameter of the target?**

To find the number of arrows, divide 1 by 10^{-2}.

$$\frac{1}{10^{-2}} = \frac{10^0}{10^{-2}}$$ $1 = 10^0$

$$= 10^{0 - \square}$$

$$= \square \text{ or } \square$$

Your Turn

a. Simplify $\frac{-8a^{-2}k^5}{18a^4k^8}$.

b. If a pencil's eraser had a diameter of 1 centimeter, how many pencil's sharpened lead tips could fit across the diameter of the eraser if each tip was 10^{-3} centimeters?

HOMEWORK ASSIGNMENT

Page(s):

Exercises:

8–4 Scientific Notation

WHAT YOU'LL LEARN

- Express numbers in scientific notation.

KEY CONCEPTS

Multiplying by Powers of 10

- If the exponent is positive, move the decimal point to the right.
- If the exponent is negative, move the decimal point to the left.

Scientific Notation A number is expressed in scientific notation when it is the form $a \times 10^n$, $1 \le a < 10$ and n is an integer.

EXAMPLES

Express each measurement in standard form.

1 8 kilobytes

8 kilobytes

$= 8 \times$ ______ The prefix *kilo-* means ______.

$=$ ______ bytes Move the decimal point ______ places right.

2 2.5 microseconds

2.5 microseconds

$= 2.5 \times$ ______ The prefix *micro-* means ______.

$=$ ______ second Move the decimal ______ places to the left.

Your Turn **Express each measurement in standard form.**

a. 5 gigabytes **b.** 4.5 milliseconds

EXAMPLES

Express each number in scientific notation.

3 325,000

$325{,}000 =$ ______ $\times 10^?$ The decimal point moves 5 places.

$=$ ______ Since 325,000 is greater than one, the exponent is positive.

4 **0.00028**

$0.00028 = \square \times 10^{?}$

The decimal point moves $\square$ places.

$= \square$

Since 0.00028 is between zero and one, the exponent is negative.

Your Turn **Express each number in scientific notation.**

a. 78,000,000

b. 0.0032

EXAMPLE

5 **Evaluate 30 × 30,000,000.**

First express each number in scientific notation. Then use the Associative and Commutative Properties to regroup factors.

$30 \times 30{,}000{,}000 = (3 \times 10^{\square})(3 \times 10^{\square})$

$= (3 \times 3)\left(\square \times \square\right)$ Associative and Commutative Properties.

$= 9 \times 10^{\square}$

$= \square$

FOLDABLES™

ORGANIZE IT

Write three large numbers and three very small numbers under the tab for Lesson 8-4. Then, write each number using scientific notation.

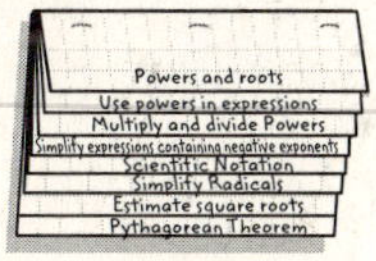

Your Turn **Evaluate each expression.**

a. $20 \times 40{,}000$

b. $\dfrac{9.6 \times 10^{9}}{4 \times 10^{5}}$

HOMEWORK ASSIGNMENT

Page(s):

Exercises:

8–5 Square Roots

WHAT YOU'LL LEARN

- Simplify radicals by using the Product and Quotient Properties of Square Roots.

BUILD YOUR VOCABULARY (pages 168–169)

The symbol, ______, called a **radical sign**, is used to indicate a ______.

An ______ that contains a square root is a **radical expression**.

A whole number that has only two ______, one and itself, is a **prime number**.

A ______ that has more than two factors is a **composite number**.

The **prime factorization** of a number is an expression of the ______ of the prime factors.

KEY CONCEPTS

Square Root A square root of a number is one of its two equal factors.

Product Property of Square Roots The square root of a product of positive numbers is equal to the product of each square root.

Quotient Property of Square Roots The square root of a quotient of positive numbers is equal to the quotient of each square root.

EXAMPLES

Simplify each expression.

1 $\sqrt{100}$

Since $10^2 =$ ______, $\sqrt{100} =$ ______.

2 $-\sqrt{81}$

Since $9^2 =$ ______, $-\sqrt{81} =$ ______.

3 $\sqrt{256}$

Find the prime factorization of 256.

$\sqrt{256} = \sqrt{2 \cdot 2 \cdot 2 \cdot 2 \cdot 2 \cdot \text{\_\_\_} \; \text{\_\_\_} \; \text{\_\_\_}}$

$= \sqrt{16} \cdot \sqrt{16}$ Use the Product Property of Square Roots.

$=$ ______ Simplify the radical.

FOLDABLES™

ORGANIZE IT

Write three radical expressions under the tab for Lesson 8-5. Then, simplify the expressions. If one cannot be simplified, explain why.

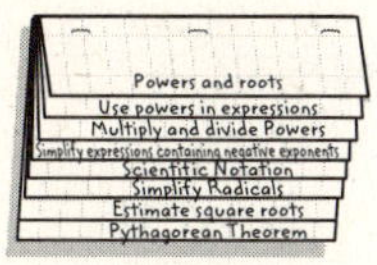

4 $\sqrt{400}$

$\sqrt{400} = \sqrt{2 \cdot 2 \cdot 2 \cdot 2 \;\_\_\_\_\; \_\_\_\_}$

$= \sqrt{16} \cdot \sqrt{25}$

$= \_\_\_\_ \cdot \_\_\_\_$ or $\_\_\_\_$

Your Turn **Simplify each expression.**

a. $\sqrt{4}$

b. $-\sqrt{121}$

c. $\sqrt{225}$

d. $\sqrt{289}$

EXAMPLE

5 **Simplify** $-\sqrt{\frac{100}{9}}$**.**

$-\sqrt{\frac{100}{9}} = -\frac{\sqrt{\_\_\_\_}}{\sqrt{\_\_\_\_}}$ Use the Quotient Property of Square Roots.

$= \_\_\_\_$

Your Turn Simplify $\sqrt{\frac{100}{324}}$.

HOMEWORK ASSIGNMENT

Page(s):

Exercises:

8–6 Estimating Square Roots

WHAT YOU'LL LEARN

- Estimate square roots.

BUILD YOUR VOCABULARY (page 168)

Irrational numbers are not rational or integers because their decimal values do not terminate or repeat.

FOLDABLES™

ORGANIZE IT

Write the square of each integer from 1–20 and write the squares under the tab for Lesson 8-6. For example, $1^2 = 1$, $2^2 = 4$, etc.

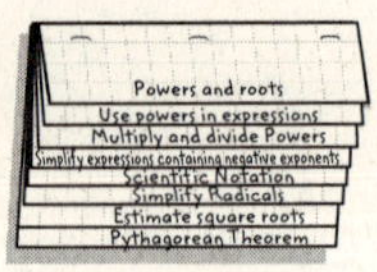

EXAMPLES

Estimate each square root.

1 $\sqrt{48}$

List some perfect squares to find the two perfect squares closest to 48.

1, 4, 9, 16, 25, 36, 49, ... 48 is between ____ and ____.

____ $< 48 <$ ____

$\sqrt{36} < \sqrt{48} < \sqrt{49}$

____ $< \sqrt{48} <$ ____

Since 48 is closer to ____ than to ____, the best whole number estimate for $\sqrt{48}$ is ____.

2 $\sqrt{200}$

____ $< 200 <$ ____ 144, 196, 225, 256, 289...

$\sqrt{196} < \sqrt{200} < \sqrt{225}$

____ $< \sqrt{200} <$ ____

Since 200 is closer to ____ than to ____, the best whole number estimate for $\sqrt{200}$ is ____.

Your Turn **Estimate each square root.**

a. $\sqrt{15}$ ____ **b.** $\sqrt{250}$ ____

HOMEWORK ASSIGNMENT

Pages(s):

Exercises:

8–7 The Pythagorean Theorem

WHAT YOU'LL LEARN

- Use the Pythagorean Theorem to solve problems.

BUILD YOUR VOCABULARY (page 168)

The side opposite the right ______ of a right triangle is called the **hypotenuse**.

The two sides that form the right angle of a ______ triangle are called the **legs**.

KEY CONCEPT

Pythagorean Theorem In a right triangle, the square of the length of the hypotenuse, *c*, is equal to the sum of the squares of the lengths of the legs, *a* and *b*.

FOLDABLES Draw a right triangle and label the right angle, hypotenuse, and the legs under the tab for Lesson 8-7. Under the triangle, write the Pythagorean Theorem.

EXAMPLE

1 Find the length of the hypotenuse of the right triangle.

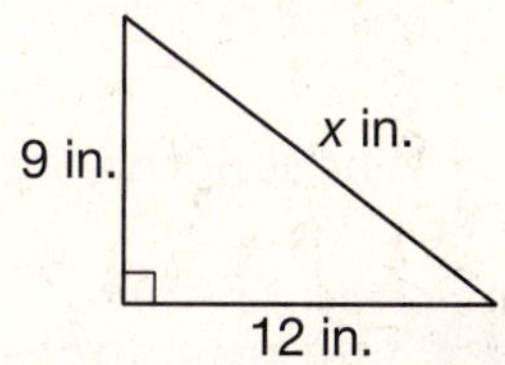

$c^2 = a^2 + b^2$	Pythagorean Theorem
$c^2 = \_\_\_\_^2 + \_\_\_\_^2$	Replace *a* and *b*.
$c^2 = \_\_\_\_ + 81$	
$c^2 = \_\_\_\_$	Find the square root of each side.
$c = \sqrt{225}$	
$c = \_\_\_\_$	

The length is ______ inches.

Your Turn Find the length of the hypotenuse of the right triangle.

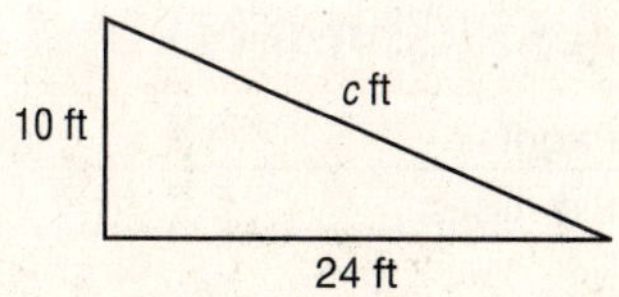

EXAMPLES

2 Find the length of one leg of a right triangle if the length of the hypotenuse is 22 centimeters and the length of the other leg is 15 centimeters. Round to the nearest tenth.

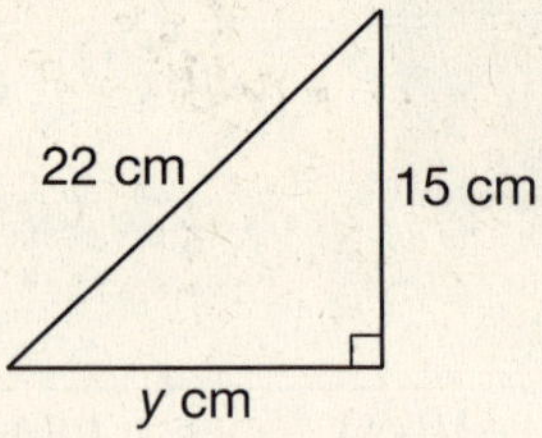

$c^2 = a^2 + b^2$	Pythagorean Theorem
$22^2 = \_\_\_\_^2 + b^2$	Replace c with 22 and a with 15.
$\_\_\_\_ = \_\_\_\_ + b^2$	
$484 - 225 = 225 - 225 + b^2$	Subtract.
$\_\_\_\_ = b^2$	
$\sqrt{259} = b$	Find the square root of each side.
$\_\_\_\_ \approx b$	Use a calculator.

The length of the leg is about ________ centimeters.

KEY CONCEPT

Converse of the Pythagorean Theorem If c is the measure of the longest side of a triangle and $c^2 = a^2 + b^2$, then the triangle is a right triangle.

3 The measures of the three sides of a triangle are 6, 11, and 13. Determine whether this triangle is a right triangle.

$c^2 = a^2 + b^2$	Pythagorean Theorem
$13^2 = 6^2 + 11^2$	Replace the variables.
$\_\_\_\_ \stackrel{?}{=} \_\_\_\_ + \_\_\_\_$	
$169 \neq 157$	

Since $c^2 \neq a^2 + b^2$, the triangle is ________ a right triangle.

Your Turn

a. Find the length of one leg of a right triangle if the length of the hypotenuse is 28 inches and the length of the other leg is 20 inches. Round to the nearest tenth.

____

b. The measures of the three sides of a triangle are 15, 18, and 21. Determine whether this triangle is a right triangle.

____

HOMEWORK ASSIGNMENT

Page(s):

Exercises:

CHAPTER 8

BRINGING IT ALL TOGETHER

STUDY GUIDE

FOLDABLES™	VOCABULARY PUZZLEMAKER	BUILD YOUR VOCABULARY
Use your **Chapter 8 Foldable** to help you study for your chapter test.	To make a crossword puzzle, word search, or jumble puzzle of the vocabulary words in Chapter 8, go to www.glencoe.com/sec/math/t_resources/free/index.php	You can use your completed **Vocabulary Builder** (pages 168–169) to help you solve the puzzle.

8-1

Powers and Exponents

1. Using 5^x, fill in the boxes with the correct terms. Five is the ______ and x is the ______.

2. Complete the table.

2^2	3^2	4^2	5^2	6^2	7^2	8^2	9^2	10^2

8-2

Multiplying and Dividing Powers

Write the letter of the correct answer at the right that best matches each expression.

3. $b^3 \cdot b^5$ ______

4. $b^3 \cdot b^3$ ______

5. $b^4 \cdot b^5 \cdot b^6$ ______

6. $\frac{b^4}{b^3}$ ______

a. b^6

b. b

c. b^9

d. b^{15}

e. b^8

8-3 Negative Exponents

7. Complete the table.

10^3	10^2	10^1	10^0	10^{-1}	10^{-2}	10^{-3}

True or False?

8. 5^{-4} is equal to $\frac{1}{5^{-4}}$.

8-4 Scientific Notation

Complete each sentence to change from scientific notation to standard notation.

9. To express 3.64×10^6 in standard notation, move the decimal point ____ places to the ____.

10. To express 7.825×10^{-3} in standard notation, move the decimal point ____ places to the ____.

Complete each sentence to change from standard notation to scientific notation.

11. To express 0.0007865 in scientific notation, move the decimal point ____ places to the right and write ____.

12. To express 54,000,000,000 in scientific notation, move the decimal point ____ places to the left and write ____.

8-5 Square Roots

Simplify each expression.

13. $-\sqrt{49}$

14. $\sqrt{121}$

15. $\sqrt{\frac{25}{81}}$

16. $-\sqrt{\frac{36}{100}}$

8-6

Estimating Square Roots

Complete the chart of the following square roots.

	Square Root	$\sqrt{?} < x < \sqrt{?}$	Estimate
	$\sqrt{45}$	$\sqrt{36} < \sqrt{45} < \sqrt{49}$	7
17.	$\sqrt{18}$		
18.	$\sqrt{88}$		
19.	$\sqrt{112}$		
20.	$\sqrt{125}$		

8-7

The Pythagorean Theorem

Complete the sentence.

21. The side opposite the right angle of a right triangle is called the ________.

22. In a right triangle, each of the two sides that form the right angle is a ________ of the right triangle.

Write an equation that you could solve to find the missing side length of each right triangle.

23.

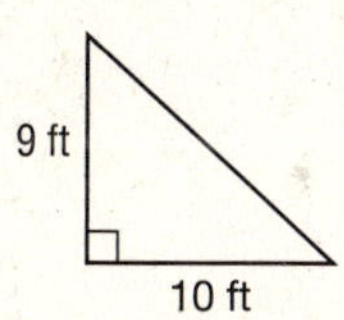

24.

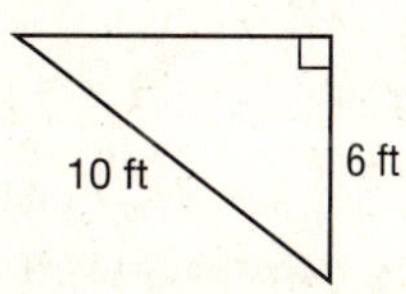

25.

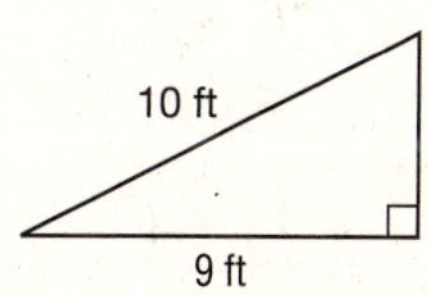

CHAPTER 8

Checklist

ARE YOU READY FOR THE CHAPTER TEST?

Math Online

Visit **algconcepts.com** to access your textbook, more examples, self-check quizzes, and practice tests to help you study the concepts in Chapter 8.

Check the one that applies. Suggestions to help you study are given with each item.

☐ **I completed the review of all or most lessons without using my notes or asking for help.**

- You are probably ready for the Chapter Test.
- You may want to take the Chapter 8 Practice Test on page 377 of your textbook as a final check.

☐ **I used my Foldable or Study Notebook to complete the review of all or most lessons.**

- You should complete the Chapter 8 Study Guide and Review on pages 374–376 of your textbook.
- If you are unsure of any concepts or skills, refer back to the specific lesson(s).
- You may also want to take the Chapter 8 Practice Test on page 377.

☐ **I asked for help from someone else to complete the review of all or most lessons.**

- You should review the examples and concepts in your Study Notebook and Chapter 8 Foldable.
- Then complete the Chapter 8 Study Guide and Review on pages 374–376 of your textbook.
- If you are unsure of any concepts or skills, refer back to the specific lesson(s).
- You may also want to take the Chapter 8 Practice Test on page 377.

Student Signature

Parent/Guardian Signature

Teacher Signature

CHAPTER 9

Polynomials

FOLDABLES™ Use the instructions below to make a Foldable to help you organize your notes as you study the chapter. You will see Foldable reminders in the margin of this Interactive Study Notebook to help you in taking notes.

Begin with a sheet of notebook paper.

STEP 1 **Fold**
Fold lengthwise to the holes.

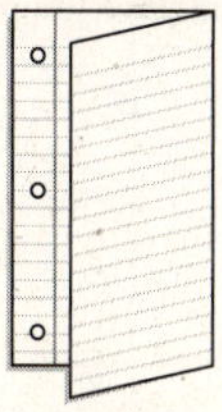

STEP 2 **Cut**
Cut along the top line and then cut three tabs.

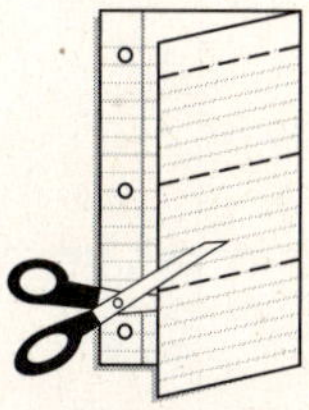

STEP 3 **Label**
Label the tabs using the lesson concepts as shown.

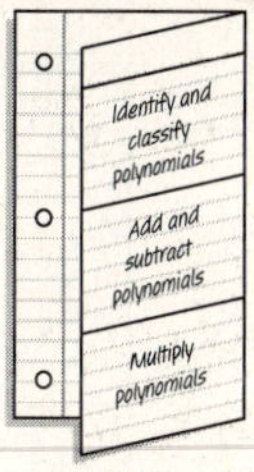

NOTE-TAKING TIP: When you take notes, be sure to write clear and concise notes so that you can study them more easily.

CHAPTER 9

BUILD YOUR VOCABULARY

This is an alphabetical list of new vocabulary terms you will learn in Chapter 9. As you complete the study notes for the chapter, you will see Build Your Vocabulary reminders to complete each term's definition or description on these pages. Remember to add the textbook page number in the second column for reference when you study.

Vocabulary Term	Found on Page	Definition	Description or Example
binomial [by-NO-mee-ul]			
degree			
FOIL method			
like terms			
monomial [ma-NO-mee-ul]			
polynomial [PA-lee-NO-mee-ul]			
trinomial [try-NO-mee-ul]			

9–1 Polynomials

What You'll Learn

- Identify and classify polynomials and find their degree.

Build Your Vocabulary (page 188)

A **monomial** is a number, a ________, or a ________ of numbers and variables that have only positive ________.

A monomial or the ________ of one or more monomials is called a **polynomial**.

A polynomial with ________ terms is a **binomial**.

A polynomial with ________ terms is a **trinomial**.

Examples

Determine whether each expression is a monomial. Explain why or why not.

1 a^2b^3c

a^2b^3c is a monomial because it is a ________ of variables.

2 $\frac{1}{x}$

$\frac{1}{x}$ is not a monomial because it includes ________.

Write It

Write three examples of expressions that are not monomials.

____

____

____

____

____

____

Your Turn **Determine whether each expression is a monomial. Explain why or why not.**

a. $-10x^2y^3z$

____

b. 5^x

____

EXAMPLES

FOLDABLES™

ORGANIZE IT

Write an example of a monomial, binomial, and trinomial under the tab for Identify and Classify Polynomials. Then write a polynomial with more than 3 terms.

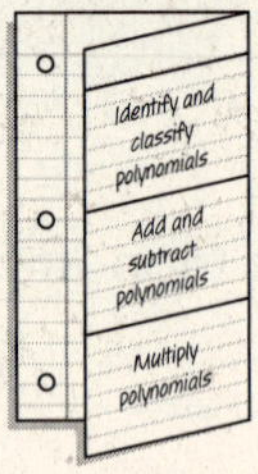

State whether each expression is a polynomial. If it is a polynomial, identify it as a monomial, binomial, or trinomial.

3 $-4x + 2$

Yes. Since the expression is the sum of ______ monomials, it is a ______.

4 $5 + 3x^2 + x + 2$

Yes. The expression $5 + 3x^2 + x + 2$ can be written as ______. Since the expression can be written as the sum of ______ monomials, it is a ______.

5 $3x^{-2} + 4x^3$

No. Since the expression contains a ______ exponent, it is not a polynomial.

Your Turn **State whether each expression is a polynomial. If it is a polynomial, identify it as a monomial, binomial, or trinomial.**

a. $9x^2 + 2x + 4$

______

b. $-3x^2 + 2x^{-3}$

______

c. $a - 3b$

______

BUILD YOUR VOCABULARY (page 188)

The **degree** of a monomial is the ______ of the ______ of the variables.

EXAMPLES

WRITE IT

Describe how a monomial can have degree zero. Write an example of a monomial with degree zero.

______

______

______

______

______

______

Find the degree of each polynomial.

6 $8b^4 + 9^2$

The degree of $8b^4 + 9^2$ is ______.

Term	Degree
$8b^4$	
9^2	

7 $2ab + 3a^2b + 5a^4b^2$

The degree of $2ab + 3a^2b + 5a^4b^2$ is ______.

Term	Degree
$2ab$	$1 + 1$ or
$3a^2b$	$2 + 1$ or
$5a^4b^2$	$4 + 2$ or

Your Turn **Find the degree of each polynomial.**

a. $3a^3b^5 + 2ab$

b. $-2x^2 - x - xy^4$

HOMEWORK ASSIGNMENT

Page(s):

Exercises:

9–2 Adding and Subtracting Polynomials

EXAMPLES

WHAT YOU'LL LEARN

- Add and subtract polynomials.

REMEMBER IT

Terms that have the same variable(s) and power(s) are like terms.

Find each sum.

1 $(3s + 4t) + (6s - 2t)$

Method 1 Group the like terms together.

$(3s + 4t) + (6s - 2t)$

$= \_\_\_\_ + (4t - 2t)$ Group the like terms.

$= \_\_\_\_ s + \_\_\_\_ t$ Distributive Property

$= \_\_\_\_$

Method 2 Add in column form.

$3s + \_\_\_\_$ Align the like terms.

$(+)\ \_\_\_\_ - 2t$

$\_\_\_\_$

2 $(b^2 + 4b - 6) + (3b^2 - 3b + 1)$

To add, use either method. Let's use Method 1. That is, group the like terms together.

$(b^2 + 4b - 6) + (3b^2 - 3b + 1)$

$= \_\_\_\_ + (4b - 3b) + \_\_\_\_$

$= \_\_\_\_ b^2 + (4 - 3)\_\_\_\_ + \_\_\_\_$

$= \_\_\_\_ b^2 + 1\_\_\_\_ + \_\_\_\_$

$= \_\_\_\_$

3 $(2d^2 + 7de - 8e^2) + (-d^2 + 8e^2)$

To add, let's use Method 2. That is, add in column form.

$$\begin{array}{lrrr} & 2d^2 & + 7de & - \square \\ (+) & -d^2 & & + 8e^2 \\ \hline & \square & & \end{array}$$

Your Turn **Find each sum.**

a. $(3x^2 + y) + (5x^2 - 2y)$

b. $(x^2 + 5x - 2) + (5x^2 - 4x + 5)$

c. $(3y^2 + 4yz - 2z^2) + (-2y^2 - 4yz + 8z^2)$

EXAMPLES

REMEMBER IT

Subtracting an integer is the same as adding its inverse.

Find each difference.

4 $(2g + 7) - (g + 2)$

Method 1 Find the additive inverse of $g + 2$. Then group the like terms together and add. The additive inverse of $g + 2$ is $-(g + 2)$ or $-g - 2$.

$(2g + 7) - (g + 2) = (2g + 7) + \square$ Add the additive inverse.

$= (2g - g) + (7 - 2)$ Group the like terms.

$= \square g + (7 - 2)$ Distributive Property

$= \square$

FOLDABLES™

ORGANIZE IT

List an example of a monomial, binomial, and trinomial under the tab for Add and Subtract Polynomials. Then, name their additive inverses.

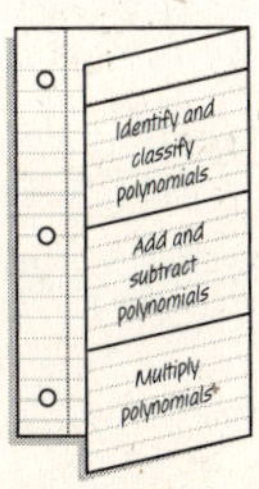

Method 2 Arrange like terms in column form.

$$\begin{array}{lr} & 2g + 7 \\ (-) & g + 2 \\ \hline & 2g + 7 \end{array}$$

Add the additive inverse.

$$\begin{array}{lrcr} & 2g & + & 7 \\ (+) & \square & - & \square \\ \hline & & \square & \end{array}$$

5 $(4a^2 - 3a + 4) - (a^2 + 6a + 1)$

The additive inverse of $(a^2 + 6a + 1)$ is $-(a^2 + 6a + 1)$.

or $-a^2 -$ ______ $-$ ______.

$(4a^2 - 3a + 4) - (a^2 + 6a + 1)$

$= (4a^2 - 3a + 4) +$ ______

$= (4a^2 - 1a^2) +$ ______ $+$ ______

$= (4 - 1)a^2 + (-3 - 6)$ ______ $+ (4 - 1)$

$=$ ______ or $3a^2 - 9a + 3$

Your Turn **Find each difference.**

a. $(6a^2 - 2) - (a^2 - 4)$

b. $(6x^2 - 2x + 2) - (x^2 + 5x + 3)$

c. $(6x^2 - x) - (10 + 5x - x^2)$

REMEMBER IT

Before adding or subtracting polynomials, be sure that the powers of the variable are in descending order.

HOMEWORK ASSIGNMENT

Pages(s):

Exercises:

9–3 Multiplying a Polynomial by a Monomial

WHAT YOU'LL LEARN

- Multiply a polynomial by a monomial.

KEY CONCEPT

Multiplying a Polynomial By a Monomial To multiply a polynomial by a monomial, use the distributive property.

EXAMPLES

Find each product.

1 $x(x + 1)$

$x(x + 1) = x$ ____ $+ x$ ____

$=$ ____ $+$ ____

2 $g(3g^2 + 4)$

$g(3g^2 + 4) = g$ ____ $+ g$ ____

$=$ ____ $+$ ____

3 $-y(2y - 6)$

$-y(2y - 6) = -y$ ____ $+ (-y)$ ____

$=$ ____ $+$ ____

4 $b^2(2b^2 - 4b - 9)$

$b^2(2b^2 - 4b - 9) = b^2$ ____ $+ b^2$ ____ $+ b^2$ ____

$=$ ____ $-$ ____ $-$ ____

Your Turn **Find each product.**

a. $-3x(x - 4)$

b. $y(5y^2 + 2)$

c. $-a(3a - 4)$

d. $x^3(3x^3 - 2x - 6)$

EXAMPLES

Solve each equation.

5 $\mathbf{3(d - 4) - 8 = 5(5d + 1) - 3}$

$3(d - 4) - 8 = 5(5d + 1) - 3$		
____ $- 12 - 8 =$ ____ $+ 5 - 3$		Distributive Property
____ $- 20 =$ ____ $+ 2$		Combine like terms.
$3d - 20 -$ ____ $= 25d + 2 -$ ____		Subtract.
____ $- 20 = 2$		
$-22d - 20 +$ ____ $= 2 +$ ____		Add ____ to each side.
$-22d =$ ____		
____ $=$ ____		Divide.

The solution is ____.

6 $\boldsymbol{a(3 + a) - 2 = a(a - 1) + 6}$

$a(3 + a) - 2 = a(a - 1) + 6$		
$3a +$ ____ $- 2 =$ ____ $- a + 6$		Distributive Property
$3a + a^2 - 2 -$ ____ $= a^2 - a + 6 -$ ____		Subtract.
____ $- 2 = -a + 6$		
$3a - 2 + a = -a + 6 + a$		Add a to each side.
____ $- 2 = 6$		
$4a - 2 +$ ____ $= 6 +$ ____		Add ____ to each side.
$4a = 8$		
____ $=$ ____		Divide.

The solution is ____.

REVIEW IT

Explain the procedure for combining exponents when multiplying like bases. (*Lesson 8-2*)

____

____

____

____

Your Turn **Solve each equation.**

a. $6(a - 3) - 2 = 5(a + 1) - 4$

b. $x(4 + x) = x(x + 10) + 6$

EXAMPLE

7 Find the area of the shaded region in simplest form.

Subtract the area of the smaller rectangle from the area of the larger rectangle.

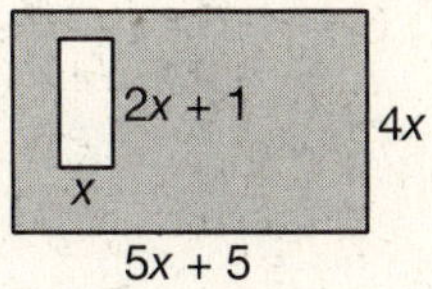

area of larger rectangle: ______ $(5x + 5)$ $A = \ell w$

area of smaller rectangle: ______ $(2x + 1)$

area of shaded region: $4x(5x + 5) -$ ______

$A = 4x(5x + 5) -$ ______

$= 4x($ ______ $) + 4x(5) - x($ ______ $) - x(1)$ Distributive Property

$=$ ______ $+ 20x -$ ______ $- x$

$=$ ______ Combine like terms.

The area is ______ square units.

Your Turn Find the area of the shaded region in simplest form.

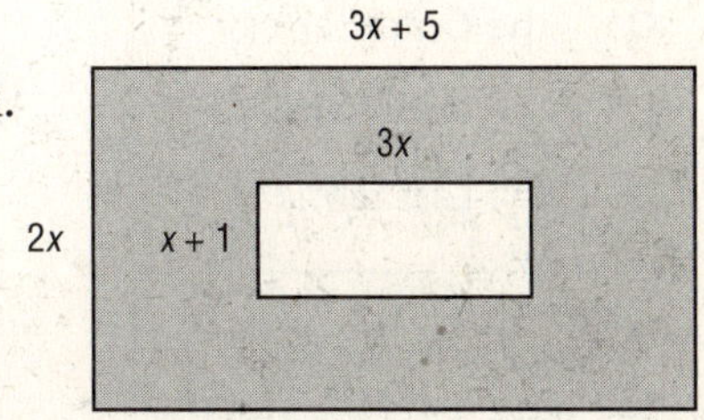

HOMEWORK ASSIGNMENT

Page(s):

Exercises:

9–4 Multiplying Binomials

WHAT YOU'LL LEARN

- Multiply two binomials.

EXAMPLE

Find the product.

1 $(x - 1)(x + 5)$

$(x - 1)(x + 5)$

$= x(x + 5) + (-1)$ ______ Distributive Property

$= x(x) + x$ ______ $+ (-1)(x) + (-1)$ ______ Distributive Property

$= x^2 +$ ______ $- x - 5$ Simplify.

$=$ ______ Combine like terms.

Your Turn **Find each product.**

a. $(x + 4)(x - 1)$

b. $(p + 3)(2p + 4)$

KEY CONCEPT

FOIL Method for Multiplying Two Binomials To multiply two binomials, find the sum of the products of

- **F** the First terms
- **O** the Outer terms
- **I** the Inner terms
- **L** the Last terms

FOLDABLES Write what the letters in FOIL stand for under the tab for Multiply Polynomials.

EXAMPLES

Find each product.

2 $(d + 2)(d + 8)$

F O I L

$(d + 2)(d + 8) = (d)$ ______ $+ (d)(8) + (2)$ ______ $+ (2)(8)$

$=$ ______ $+ 8d +$ ______ $+ 16$

$=$ ______ Combine like terms.

3 **$(e + 4)(2e - 4)$**

F O I L

$(e + 4)(2e - 4) = (e)$ ☐ $+ (e)(-4) + (4)$ ☐ $+ (4)(-4)$

$=$ ☐ $- 4e +$ ☐ $- 16$

$=$ ☐ Combine like terms.

4 **$(5x + y)(4x - 2y)$**

F O I L

$(5x + y)(4x - 2y) = (5x)$ ☐ $+ (5x)(-2y) + (y)$ ☐ $+ (y)(-2y)$

$=$ ☐ $- 10xy +$ ☐ $- 2y^2$

$=$ ☐ Combine like terms.

EXAMPLE

5 **Find the product of $a - 2$ and $a^2 - 4$.**

$(a - 2)(a^2 - 4) = (a)(a^2) + (a)$ ☐ $+ (-2)(a^2) + (-2)$ ☐

$= a^3 -$ ☐ $- 2a^2 +$ ☐ There are no like terms.

$=$ ☐

Your Turn **Find each product.**

a. $(y + 9)(y + 5)$

b. $(k + 3)(3k - 5)$

c. $(2x + y)(x - 3y)$

d. $(y + 2)(y^2 + 5)$

HOMEWORK ASSIGNMENT

Page(s):

Exercises:

9–5 Special Products

What You'll Learn

- Develop and use the patterns for $(a + b)^2$, $(a - b)^2$, and $(a + b)(a - b)$.

Key Concepts

Square of a Sum and Square of a Difference

$(a + b)^2 = a^2 + 2ab + b^2$

$(a - b)^2 = a^2 - 2ab + b^2$

EXAMPLES

Find each product.

1 $(b + 5)^2$

$(a + b)^2 = \square + 2ab + \square$ — Square of a Sum

$(b + 5)^2 = \square + 2(b)(5) + \square$ — Replace a with $\square$ and b with $\square$.

$= \square$

2 $(c - 3)^2$

$(a - b)^2 = \square - 2ab + \square$ — Square of a Difference

$(c - 3)^2 = \square - 2(c)(3) + \square$ — Replace a with $\square$ and b with $\square$.

$= \square$

3 $(2d + 1)^2$

$(a + b)^2 = \square + 2ab + \square$ — Square of a Sum

$(2d + 1)^2 = \square + 2(2d)(1) + \square$ — Replace a with $2d$ and b with 1.

$= \square$

4 $(3e - 3f)^2$

$(a - b)^2 = \square - 2ab + \square$ — Square of a Difference

$(3e - 3f)^2 = \square - 2(3e)(3f) + \square$ — Replace a with $3e$ and b with $3f$.

$= \square$

EXAMPLES

Find each product.

5 $(3 + a)(3 - a)$

$(a + b)(a - b) =$ ______ $- b^2$ Product of a Sum and a Difference

$(3 + a)(3 - a) =$ ______ $- a^2$ Replace *a* with ______ and *b* with ______.

$=$ ______

6 $(5b - 2)(5b + 2)$

$(a - b)(a + b) =$ ______ $- b^2$ Product of a Sum and a Difference

$(5b - 2)(5b + 2) =$ ______ $- (2)^2$ Replace ______ with $5b$ and *b* with ______.

$=$ ______

Your Turn **Find each product.**

a. $(y + 3)^2$

______

b. $(k - 4)^2$

______

c. $(2x + 5)^2$

______

d. $(3x - 2y)^2$

______

e. $(x + 6)(x - 6)$

______

f. $(4x - 3y)(4x + 3y)$

______

FOLDABLES™

ORGANIZE IT

Write the three special product models under the tab for Multiply Polynomials. Then, write an example of each and include a model.

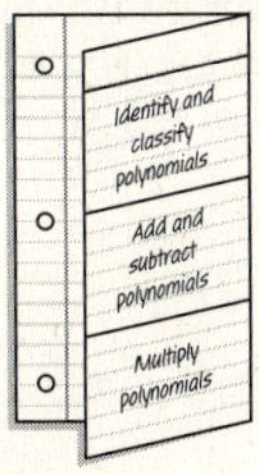

KEY CONCEPT

Product of a Sum and a Difference

$(a + b)(a - b) = a^2 - b^2$

HOMEWORK ASSIGNMENT

Page(s):

Exercises:

CHAPTER 9

BRINGING IT ALL TOGETHER

STUDY GUIDE

FOLDABLES	VOCABULARY PUZZLEMAKER	BUILD YOUR VOCABULARY
Use your **Chapter 9 Foldable** to help you study for your chapter test.	To make a crossword puzzle, word search, or jumble puzzle of the vocabulary words in Chapter 9, go to www.glencoe.com/sec/math/t_resources/free/index.php	You can use your completed **Vocabulary Builder** (page 188) to help you solve the puzzle.

9-1 Polynomials

Match the expression on the left with the correct term at the right.

1. $-5x^2y^3z - 2x$
2. $\frac{3x}{y}$
3. $12a^2b^3c^4$
4. $2k + 3m + 4mn$

a. monomial
b. binomial
c. trinomial
d. not a polynomial

Find the degree of each polynomial.

5. $5ab + 2a^2b^2 + 4a^2b$
6. $-6xy^3z + xy$
7. 12
8. $3a^2b - 2a^2$

9-2 Adding and Subtracting Polynomials

Find each sum or difference.

9. $(-3x^3 + 4x^2 + 5x + 1) + (-5x^3 - 2x^2 + 2x - 7)$

10. $(5k + 4) - (k + 1)$

9-3 Multiplying a Polynomial by a Monomial

11. The ______ property is used to multiply a polynomial by a monomial.

Find each product.

12. $2y^2(3y^2 + 2y - 7)$

13. $-3x^3(x^3 - 2x^2 + 3)$

9-4 Multiplying Binomials

14. Complete. To multiply two binomials, find the sum of the products of

F = ______,

O = ______,

I = ______, and

L = ______.

9-5 Special Products

Match each model to its special product name.

15. $(a - b)^2 = a^2 - 2ab + b^2$ ______

16. $(a + b)^2 = a^2 + 2ab + b^2$ ______

17. $(a + b)(a - b) = a^2 - b^2$ ______

a. square of a sum

b. square of a difference

c. product of a sum and a difference

d. sum of a difference

CHAPTER 9

Checklist

ARE YOU READY FOR THE CHAPTER TEST?

Math Online

Visit **algconcepts.com** to access your textbook, more examples, self-check quizzes, and practice tests to help you study the concepts in Chapter 9.

Check the one that applies. Suggestions to help you study are given with each item.

☐ **I completed the review of all or most lessons without using my notes or asking for help.**

- You are probably ready for the Chapter Test.
- You may want to take the Chapter 9 Practice Test on page 415 of your textbook as a final check.

☐ **I used my Foldable or Study Notebook to complete the review of all or most lessons.**

- You should complete the Chapter 9 Study Guide and Review on pages 412–414 of your textbook.
- If you are unsure of any concepts or skills, refer back to the specific lesson(s).
- You may also want to take the Chapter 9 Practice Test on page 415.

☐ **I asked for help from someone else to complete the review of all or most lessons.**

- You should review the examples and concepts in your Study Notebook and Chapter 9 Foldable.
- Then complete the Chapter 9 Study Guide and Review on pages 412–414 of your textbook.
- If you are unsure of any concepts or skills, refer back to the specific lesson(s).
- You may also want to take the Chapter 9 Practice Test on page 415.

Student Signature

Parent/Guardian Signature

Teacher Signature

CHAPTER 10

Factoring

FOLDABLES™ Use the instructions below to make a Foldable to help you organize your notes as you study the chapter. You will see Foldable reminders in the margin of this Interactive Study Notebook to help you in taking notes.

Begin with a sheet of plain $8\frac{1}{2}$" by 11" paper.

STEP 1 **Fold**
Fold in half lengthwise.

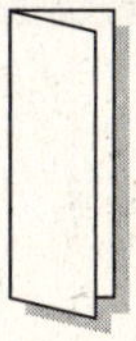

STEP 2 **Fold**
Fold again in thirds.

STEP 3 **Open**
Cut along the second fold to make three tabs.

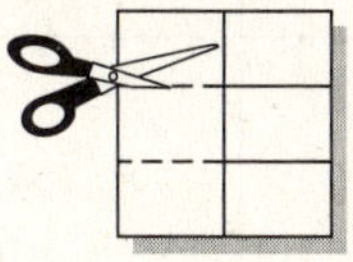

STEP 4 **Label**
Label each tab as shown.

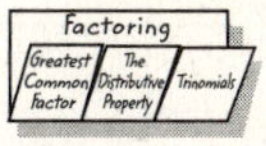

NOTE-TAKING TIP: When you start a new lesson, write down the objectives. Then look for sample problems and solutions that illustrate the objectives.

CHAPTER 10

BUILD YOUR VOCABULARY

This is an alphabetical list of new vocabulary terms you will learn in Chapter 10. As you complete the study notes for the chapter, you will see Build Your Vocabulary reminders to complete each term's definition or description on these pages. Remember to add the textbook page number in the second column for reference when you study.

Vocabulary Term	Found on page	Definition	Description or Example
difference of squares			
factoring			
greatest common factor (GCF)			
perfect square trinomial			
prime polynomial			

10–1 Factors

WHAT YOU'LL LEARN

- Find the greatest common factor of a set of numbers or monomials.

EXAMPLES

Find the factors of each number. Then classify each number as prime or composite.

1 **47**

There is only one pair of whole numbers whose product is 47.

$1 \times$ ______

The factors of 47 are ______ and ______. Therefore, 47 is a ______ number.

2 **35**

To find the factors of 35, list all pairs of whole numbers whose product is 35.

$1 \times$ ______ $\quad 5 \times$ ______

The factors of 35 are 1, 5, ______, and ______. Since 35 has more than ______ factors, it is a ______ number.

Your Turn **Find the factors of each number. Then classify each number as prime or composite.**

a. 51

b. 18

REVIEW IT

How can you tell that the expression listed in Example 3 is a monomial? (*Lesson 9-1*)

______

______

______

______

______

______

EXAMPLES

Factor each monomial.

3 $\mathbf{16b^2c^2}$

$16b^2c^2 = 2 \cdot 2 \cdot 2 \cdot$ ______ $\cdot\, b \,\cdot$ ______ $\cdot\, c \cdot c$

4 $-15xy^2$

To factor a negative integer, first express it as the product of a whole number and -1. Then find the prime factorization.

$-15xy^2 = \_\_\_\_ \cdot 15xy^2$

$-15xy^2 = -1 \cdot 3 \cdot \_\_\_\_ \cdot x \cdot y \cdot \_\_\_\_$

Your Turn **Factor each monomial.**

a. $-36xy^2$ **b.** $6a^3b^2$

EXAMPLES

Find the GCF of each set of numbers or monomials.

5 **12, 20, and 24**

$12 = (2) \cdot (2) \cdot 3$ Find the prime factorization of each number.

$20 = (2) \cdot (2) \cdot 5$ Line up as many factors as possible.

$24 = (2) \cdot (2) \cdot 2 \cdot 3$ Circle the common factors.

The GCF of 12, 20, and 24 is ____ $\cdot$ ____ or ____.

6 $\mathbf{21ab^2}$ **and** $\mathbf{9a^2b}$

$21ab^2 = (3) \cdot 7 \cdot (a) \cdot (b) \cdot b$

$9a^2b = (3) \cdot 3 \cdot (a) \cdot a \cdot (b)$

The GCF of $21ab^2$ and $9a^2b$ is ____ $\cdot$ ____ $\cdot$ ____ or ____.

Your Turn **Find the GCF of each set of numbers or monomials.**

a. 10, 20, and 50 **b.** 10 and 11 **c.** $9a^2b$ and $12a^3b^2$

KEY CONCEPT

Greatest Common Factor The greatest common factor of two or more integers is the product of the prime factors common to the integers.

FOLDABLES™ Find the greatest common factor for a set of three numbers and write it under the tab for Greatest Common Factor.

HOMEWORK ASSIGNMENT

Page(s):

Exercises:

10–2 Factoring Using the Distributive Property

WHAT YOU'LL LEARN

- Use the GCF and the Distributive Property to factor polynomials.

BUILD YOUR VOCABULARY (page 206)

The process of finding the factors of a product is **factoring**.

A polynomial that cannot be factored is called a **prime polynomial**.

EXAMPLES

Factor each polynomial.

1 $24y + 18y^2$

First, find the GCF of $24y$ and $18y^2$.

$24y = \boxed{2} \cdot 2 \cdot 2 \cdot \boxed{3} \cdot \boxed{y}$

$18y^2 = \boxed{2} \cdot \boxed{3} \cdot 3 \cdot \boxed{y} \cdot y$

The GCF of $24y$ and $18y^2$ is ______. Write each term as a product of the GCF and its remaining factors.

$24y + 18y^2 =$ ______$(4) +$ ______$(3y)$

$= 6y$ ______ Distributive Property

2 $18fg - 21gh^2$

First, find the GCF of $18fg$ and $21gh^2$.

$18fg = 2 \cdot \boxed{3} \cdot 3 \cdot f \cdot \boxed{g}$

$21gh^2 = \boxed{3} \cdot 7 \cdot \boxed{g} \cdot h \cdot h$

The GCF is ______.

$18fg - 21gh^2 =$ ______$(6f) -$ ______$(7h^2)$

$= 3g($ ______ $)$ Distributive Property

REVIEW IT

How can you use the Distributive Property to check your work? (*Lesson 9-3*)

______

______

______

______

______

______

Your Turn **Factor each polynomial.**

a. $10a^2 + 5a$ ______

b. $12mn^2 - 15n$ ______

EXAMPLE

3 Factor $5a + 20ab + 10a^2$.

$$5a = \boxed{5} \cdot \boxed{a}$$
$$20ab = 2 \cdot 2 \cdot \boxed{5} \cdot \boxed{a} \cdot b$$
$$10a^2 = 2 \cdot \boxed{5} \cdot \boxed{a} \cdot a$$

The GCF is ______. When $5a$ is factored from $5a$, the remaining factor is ______.

$5a + 20ab + 10a^2 =$ ______$(1) +$ ______$(4b) +$ ______$(2a)$

$= 5a$______ Distributive Property

Your Turn **Factor each polynomial.**

a. $6ab + 15ab^2 + 3a^2b^2$

b. $15c + 11ab^2$

EXAMPLE

4 Divide $(24a^2 - 20a)$ by $4a$.

$(24a^2 - 20a) \div 4a = \dfrac{24a^2}{\square} - \dfrac{20a}{\square}$ Divide each term by ______.

$= \dfrac{\overset{1}{\cancel{2}} \cdot \overset{1}{\cancel{2}} \cdot 2 \cdot 3 \cdot \overset{1}{\cancel{a}} \cdot a}{\underset{1}{\cancel{2}} \cdot \underset{1}{\cancel{2}} \cdot \underset{1}{\cancel{a}}} - \dfrac{\overset{1}{\cancel{2}} \cdot \overset{1}{\cancel{2}} \cdot 5 \cdot \overset{1}{\cancel{a}}}{\underset{1}{\cancel{2}} \cdot \underset{1}{\cancel{2}} \cdot \underset{1}{\cancel{a}}}$ Simplify.

$=$ ______ $-$ ______

Therefore, $(24a^2 - 20a) \div 4a =$ ______.

FOLDABLES™

ORGANIZE IT

Multiply a monomial and a trinomial together under the tab for Distributive Property.

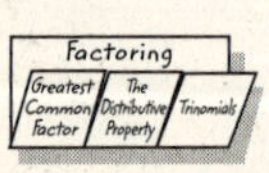

Your Turn Divide $(36ab^2 - 9a^2b)$ by $9ab$.

HOMEWORK ASSIGNMENT

Page(s):

Exercises:

10–3 Factoring Trinomials: $x^2 + bx + c$

EXAMPLES

WHAT YOU'LL LEARN

- Factor trinomials of the form $x^2 + bx + c$.

REMEMBER IT

The product of a positive integer and a negative integer is negative. The product of two negative integers is positive.

Factor each trinomial.

1 $x^2 + x - 12$

$$x^2 + x - 12 = (x + \square)(x + \square)$$

Find integers whose product is ______ and whose sum is ______.

Product	Integers	Sum
−12	−2, 6	−2 + 6 = ______
−12	2, −6	2 + (−6) = ______
−12	−3, 4	−3 + 4 = ______

You can stop listing factors when you find a pair that works.

Therefore, $x^2 + x - 12 = (x - \text{______})(x + \text{______})$.

2 $x^2 - 9x + 8$

$$x^2 - 9x + 8 = (x + \square)(x + \square)$$

Find integers whose product is ______ and whose sum is ______.

Product	Integers	Sum
8	−1, −8	−1 + (−8) = ______

Therefore, $x^2 - 9x + 8 = (x - \text{______})(x - \text{______})$.

REMEMBER IT Square models can be used to help find the correct combination when factoring trinomials.

3 $x^2 + x + 1$

$x^2 + x + 1 = (x + \square)(x + \square)$

Find integers whose product is ☐ and whose sum is ☐.

Product	Integers	Sum
1	1, 1	$1 + 1 =$ ☐
1	$-1, -1$	$-1 + (-1) =$ ☐

There are no factors of ☐ whose sum is ☐.

Therefore, $x^2 + x + 1$ is a ☐ polynomial.

Your Turn **Factor each trinomial.**

a. $x^2 + 14x + 24$

☐

b. $x^2 + 7x - 18$

☐

c. $x^2 + 2x - 1$

☐

EXAMPLE

4 **Factor $4x^2 - 8x - 60$.**

First, check for a GCF.

$4x^2 - 8x - 60 =$ ☐ $(x^2 - 2x - 15)$ The GCF is ☐.

Now, factor $x^2 - 2x - 15$.

$x^2 - 2x - 15 = \left(x + \square\right)\left(x - \square\right)$

$4x^2 - 8x - 60 =$ ☐

Your Turn Factor $5x^2 - 10x - 15$.

☐

HOMEWORK ASSIGNMENT

Page(s):

Exercises:

10–4 Factoring Trinomials: $ax^2 + bx + c$

EXAMPLE

What You'll Learn

- Factor trinomials of the form $ax^2 + bx + c$.

Foldables

Organize It

Choose two binomials and use FOIL to multiply them together under the tab for Trinomials. Then use the factoring techniques you've learned to factor the result back into two binomials.

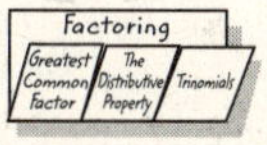

1 Factor $3y^2 + 7y - 6$.

$3y^2$ is the product of the first terms, and -6 is the product of the last terms.

$$3y^2 + 7y - 6 = (3y + \square)(y + \square)$$

The last term, ______, is negative. The sum of the inside and outside terms, ______, is positive. So, one factor must be ______ and one must be ______. Try factor pairs of ______ until the sum of the products of the Outer and Inner terms is ______.

Try 2 and -3.

$(3y + \_\_\_)(y - \_\_\_) = 3y^2 - 9y + 2y - 6$

$= 3y^2$ ______ $- 6$ ______ is not the correct middle term.

$(3y - \_\_\_)(y + \_\_\_) = 3y^2 + 6y - 3y - 6$

$= 3y^2$ ______ $- 6$ ______ is not the correct middle term.

Try -2 and 3.

$(3y - \_\_\_)(y + \_\_\_) = 3y^2 + 9y - 2y - 6$

$= 3y^2 +$ ______ $- 6$

Therefore, $3y^2 + 7y - 6 =$ ______.

Your Turn **Factor each trinomial.**

a. $2x^2 - x - 3$

b. $5z^2 + 22z + 8$

EXAMPLE

2 **Factor $4x^2 - 4x - 15$.**

Number	Factor Pairs
4	4 and 1, 2 and 2
−15	3 and −5, −3 and 5, 15 and −1, −15 and 1

Try 4 and 1.

$(4x + 3)(x - 5) = 4x^2 - \square + \square - 15$

$= 4x^2 - \square - 15$ $\square$ is not the correct middle term.

$(4x - 15)(x + 1) = 4x^2 + \square - \square - 15$

$= 4x^2 - \square - 15$ $\square$ is not the correct middle term.

Try 2 and 2. $(2x + 3)(2x - 5) = 4x^2 - \square + \square - 15$

$= 4x^2 - \square - 15$

Therefore, $4x^2 - 4x - 15 = \square$.

HOMEWORK ASSIGNMENT

Page(s):

Exercises:

Your Turn Factor $4c^2 - 16c + 7$.

10–5 Special Factors

WHAT YOU'LL LEARN

- Recognize and factor the differences of squares and perfect square trinomials.

KEY CONCEPT

Factoring Perfect Square Trinomials A perfect square trinomial is a trinomial that has two equal binomial factors. For example, $x^2 + 6x + 9 = (x + 3)(x + 3)$ and $x^2 - 6x + 9 = (x - 3)(x - 3)$.

REMEMBER IT

In a perfect square trinomial, the last term is always positive.

EXAMPLES

Determine whether each trinomial is a perfect square trinomial. If so, factor it.

1 $x^2 + 14x + 49$

To determine whether $x^2 + 14x + 49$ is a perfect square trinomial, answer each question.

- Is the first term a perfect square?

  ______ x^2 is the square of ______.

- Is the last term a perfect square?

  ______ 49 is the square of ______.

- Is the middle term twice the product of x and ______?

  ______ $14x =$ ______$(7x)$.

Therefore, $x^2 + 14x + 49$ is a perfect square trinomial.

$x^2 + 14x + 49 =$ ______.

2 $9a^2 + 16a + 4$

- Is the first term a perfect square?

  ______ $9a^2$ is the square of ______.

- Is the last term a perfect square?

  ______ 4 is the square of ______.

- Is the middle term twice the product of $3a$ and ______?

  ______ $2($______$) \neq 16a$

Therefore, $9a^2 + 16a + 4$ is not a perfect square trinomial.

Your Turn **Determine whether each trinomial is a perfect square trinomial. If so, factor it.**

a. $x^2 - 10x + 25$

b. $4c^2 - 20c + 100$

c. $9m^2 + 12m + 4$

EXAMPLES

KEY CONCEPT

Factoring a Difference of Squares A binomial that can be factored into two binomials is a difference of squares. For example, $x^2 - 9 = (x - 3)(x + 3)$.

Determine whether each binomial is the difference of squares. If so, factor it.

3 $d^2 - 81$

d^2 and 81 are both perfect ______ and $d^2 - 81$ is a ______.

$d^2 - 81 = (\ \_\_\ )^2 - (\ \_\_\ )^2$ ______ $= d^2$, ______ $= 81$

$= (d - 9)(\ \_\_\_\_\ )$ Difference of Squares

4 $f^2 + 64$

f^2 and 64 are both perfect squares. But $f^2 + 64$ is a ______, not a difference. Therefore $f^2 + 64$ is not a difference of squares. It is a ______ polynomial.

Your Turn **Determine whether each binomial is the difference of squares. If so, factor it.**

a. $r^2 - 100$

b. $n^2 + 4$

c. $9x^2 - 121$

HOMEWORK ASSIGNMENT

Page(s):

Exercises:

CHAPTER 10

BRINGING IT ALL TOGETHER

STUDY GUIDE

FOLDABLES	VOCABULARY PUZZLEMAKER	BUILD YOUR VOCABULARY
Use your **Chapter 10 Foldable** to help you study for your chapter test.	To make a crossword puzzle, word search, or jumble puzzle of the vocabulary words in Chapter 10, go to www.glencoe.com/sec/math/t_resources/free/index.php	You can use your completed **Vocabulary Builder** (page 206) to help you solve the puzzle.

10-1

Factors

Find the prime factorization of each number or monomial.

1. 36

2. 14

3. $25a^3b$

4. $27x$

5. 81

6. $42mn^2$

10-2

Factoring Using the Distributive Property

Refer to the polynomial $5ab + 25b^2 - 5b$.

7. What is the GCF of this polynomial?

8. Write this polynomial in factored form.

9. Explain how to check your answer.

10-3

Factoring Trinomials: $x^2 + bx + c$

Tell what sum and product you want m and n to have to use the pattern $(x + m)(x + n)$ to factor the given trinomial.

10. $x^2 + 10x + 24$ sum: ____ product: ____

11. $x^2 - 12x + 20$ sum: ____ product: ____

12. $x^2 - 4x - 21$ sum: ____ product: ____

13. $x^2 + 6x - 16$ sum: ____ product: ____

Match each trinomial with the correct factored form. If the trinomial will not factor, choose answer F.

14. $x^2 + 7x - 30$ ____

15. $x^2 - 7x + 6$ ____

16. $x^2 + 8x + 15$ ____

17. $x^2 - 6x - 16$ ____

18. $x^2 + 2x - 5$ ____

a. $(x + 2)(x - 8)$
b. $(x - 1)(x - 6)$
c. $(x + 10)(x - 3)$
d. $(x + 3)(x + 5)$
e. $(x + 2)(x - 4)$
f. prime polynomial

10-4

Factoring Trinomials: $ax^2 + bx + c$

Refer to the trinomial $4x^2 + 11x + 6$.

19. What are the possibilities for the first term in each binomial?

20. What are the possibilities for the last term in each binomial?

10-5

Special Factors

Explain why each binomial is a difference of squares.

21. $4x^2 - 25$

22. $49a^2 - 64b^2$

Match each polynomial from the first column with a factoring technique in the second column. Some of the techniques may be used more than once. If none of the techniques can be used to factor the polynomial, choose *prime polynomial*.

23. $9x^2 - 64$

24. $9x^2 + 12x + 4$

25. $x^2 - 5x + 6$

26. $4x^2 + 13x + 9$

27. $x^2 + 25$

28. $x^2 - 4x + 4$

29. $2x^2 - 16$

a. factor as $x^2 + bx + c$

b. factor as $ax^2 + bx + c$

c. difference of squares

d. perfect square trinomial

e. factor out the GCF

f. prime polynomial

CHAPTER 10 Checklist

ARE YOU READY FOR THE CHAPTER TEST?

Math Online

Visit **algconcepts.com** to access your textbook, more examples, self-check quizzes, and practice tests to help you study the concepts in Chapter 10.

Check the one that applies. Suggestions to help you study are given with each item.

☐ **I completed the review of all or most lessons without using my notes or asking for help.**

- You are probably ready for the Chapter Test.
- You may want to take the Chapter 10 Practice Test on page 453 of your textbook as a final check.

☐ **I used my Foldable or Study Notebook to complete the review of all or most lessons.**

- You should complete the Chapter 10 Study Guide and Review on pages 450–452 of your textbook.
- If you are unsure of any concepts or skills, refer back to the specific lesson(s).
- You may also want to take the Chapter 10 Practice Test on page 453.

☐ **I asked for help from someone else to complete the review of all or most lessons.**

- You should review the examples and concepts in your Study Notebook and Chapter 10 Foldable.
- Then complete the Chapter 10 Study Guide and Review on pages 450–452 of your textbook.
- If you are unsure of any concepts or skills, refer back to the specific lesson(s).
- You may also want to take the Chapter 10 Practice Test on page 453.

Student Signature

Parent/Guardian Signature

Teacher Signature

CHAPTER 11

Quadratic and Exponential Functions

FOLDABLES™ Use the instructions below to make a Foldable to help you organize your notes as you study the chapter. You will see Foldable reminders in the margin of this Interactive Study Notebook to help you in taking note.

Begin with four sheets of plain paper.

STEP 1 **Fold**
Fold each sheet in half along the width.

STEP 2 **Unfold**
Unfold each sheet and tape to form one long piece.

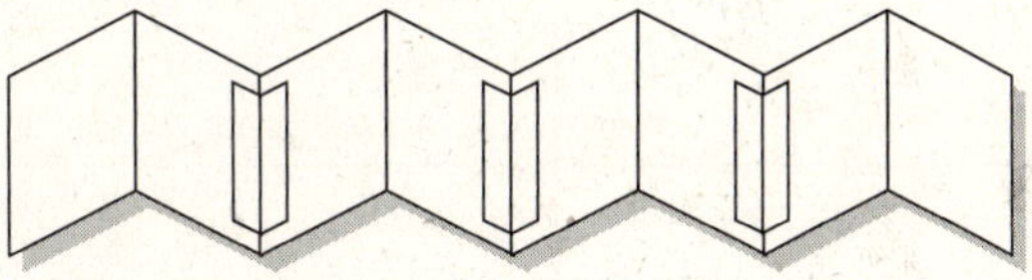

STEP 3 **Label**
Label each page with the lesson number as shown.

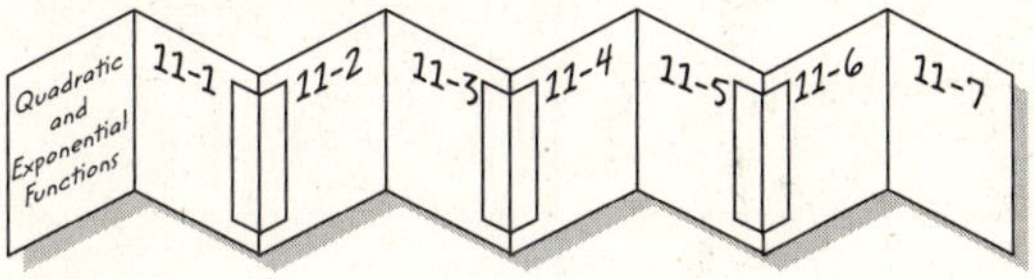

STEP 4 **Refold**
Refold to form a booklet. Label the front cover "Quadratic and Exponential Functions."

NOTE-TAKING TIP: When taking notes, listen for key words your teacher may use to emphasize important concepts.

CHAPTER 11

BUILD YOUR VOCABULARY

This is an alphabetical list of new vocabulary terms you will learn in Chapter 11. As you complete the study notes for the chapter, you will see Build Your Vocabulary reminders to complete each term's definition or description on these pages. Remember to add the textbook page number in the second column for reference when you study.

Vocabulary Term	Found on Page	Definition	Description or Example
axis of symmetry [SIH-muh-tree]			
completing the square			
discriminant [dis-KRIMH-uh-nunt]			
exponential function [EKS-po-NEN-chul]			
geometric sequence [JEE-uh-MET-rik]			
initial value			
maximum			

Vocabulary Term	Found on Page	Definition	Description or Example
minimum			
parabola [puh-RA-buh-la]			
quadratic equation [kwad-RAT-ik]			
Quadratic Formula			
quadratic function			
roots			
vertex [VER-teks]			
zeros			

11–1 Graphing Quadratic Functions

What You'll Learn

- Graph quadratic functions.

Build Your Vocabulary (page 223)

The shape of a ______ function is called a **parabola**.

Key Concept

Quadratic Function A quadratic function is a function that can be described by an equation of the form $y = ax^2 + bx + c$, where $a \neq 0$.

Examples

1 Graph $y = x^2 - 2$ by making a table of values.

First choose ______ values for x. Evaluate the function for each x-value.

x	$x^2 - 2$	y	(x, y)
-2	$(-2)^2 - 2$		
-1	$(-1)^2 - 2$		
0	$0^2 - 2$		
1	$1^2 - 2$		
2	$2^2 - 2$		

Graph the points and connect them with a ______ curve.

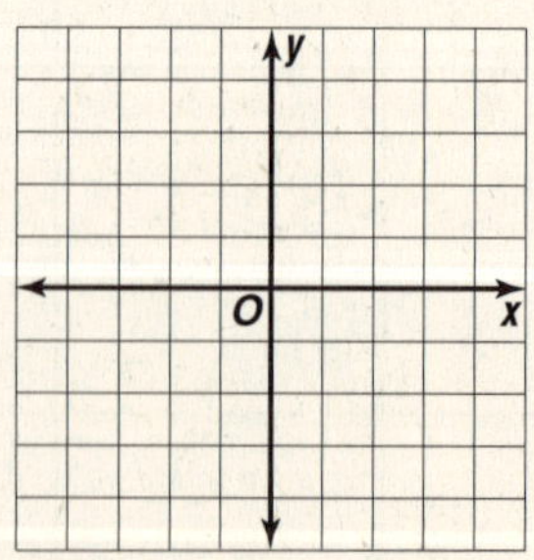

Your Turn **Graph each quadratic equation by making a table of values.**

a. $y = -x^2 - 3$

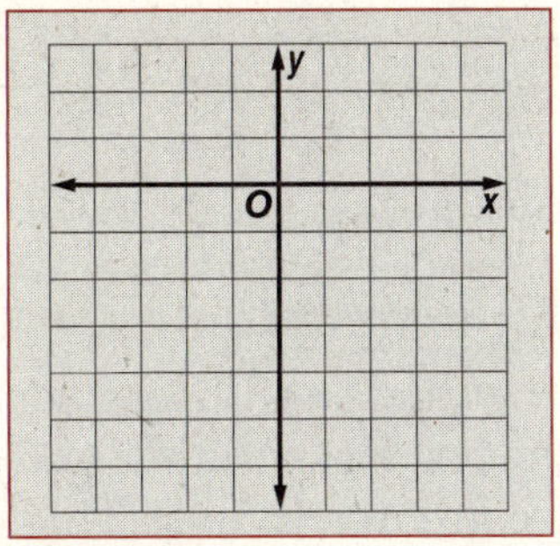

b. $y = x^2 + 2$

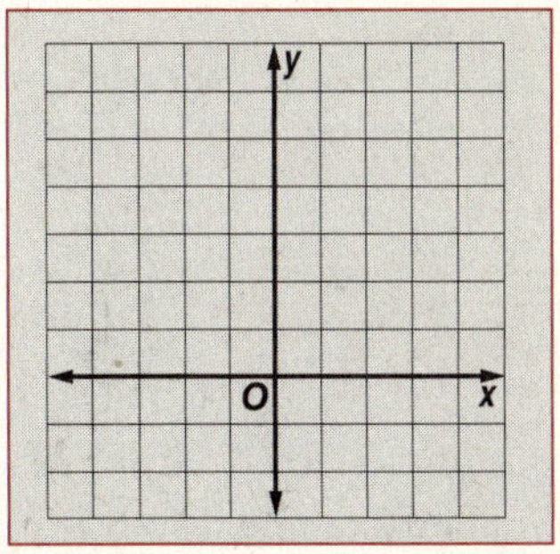

EXAMPLE

2 Use characteristics of quadratic functions to graph $y = -x^2 + 2x + 1$.

KEY CONCEPT

Equation of the Axis of Symmetry The equation of the axis of symmetry for the graph of $y = ax^2 + bx + c$, where $a \neq 0$, is $x = -\frac{b}{2a}$.

FOLDABLES Draw a parabola under the tab for Lesson 11-1. Label the axis of symmetry and the vertex.

a. Find the equation of the axis of symmetry.

$y = ax^2 + bx + c$

↓ ↓ ↓

$y = -x^2 + 2x + 1$ $\quad -x^2 = -1x^2$

So, $a =$ ______, $b =$ ______, and $c =$ ______.

Now, find the equation of the axis of symmetry.

$x =$ ______ Equation of axis symmetry

$x = -\frac{\text{\_\_\_\_\_\_}}{2(\text{\_\_\_\_\_\_})}$ $\quad a =$ ______, $b =$ ______

$x =$ ______ Simplify.

b. Find the coordinates of the vertex of the parabola.

Since the equation of the axis of symmetry is ______,

the x-coordinate of the vertex must be ______. Substitute

______ for x in the equation $y = -x^2 + 2x + 1$ to solve for y.

$y = -x^2 + 2x + 1$

$= -\_\_\_\_^2 + 2\_\_\_\_ + 1$

$= \_\_\_\_ + \_\_\_\_ + 1$ or $\_\_\_\_$

The point at ______ is the vertex.

REMEMBER IT

The axis of symmetry can help you graph quadratic functions. If you know the axis of symmetry, additional points can be found by reflecting across the axis of symmetry.

c. Graph the function.

Construct a table. Choose some values for x that are less than 1 and some that are greater than 1. This ensures that points on each side of the axis of symmetry are graphed.

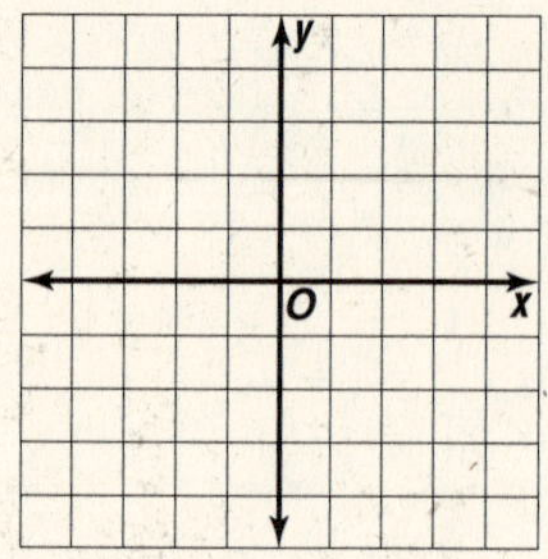

x	$-x^2 + 2x + 1$	y	(x, y)
-1	$-(-1)^2 + 2(-1) + 1$		
0	$-(0)^2 + 2(0) + 1$		
1	$-(1)^2 + 2(1) + 1$		
2	$-(2)^2 + 2(2) + 1$		
3	$-(3)^2 + 2(3) + 1$		

Your Turn **Use characteristics of quadratic functions to graph $y = x^2 - 4x - 3$.**

a. Find the equation of the axis of symmetry.

b. Find the coordinates of the vertex of the parabola.

c. Graph the function.

HOMEWORK ASSIGNMENT

Page(s):

Exercises:

11–2 Families of Quadratic Functions

What You'll Learn

- Learn the characteristics of families of parabolas.

Key Concept

Families of Parabolas In families of parabolas, graphs either share a vertex or an axis of symmetry, or both. Also, a family can consist of parabolas of the same shape.

Foldables Draw an example of a family of parabolas that share the same axis of symmetry under the tab for Lesson 11-2.

EXAMPLES

Graph each group of equations on the same screen. Compare and contrast the graphs. What conclusions can be drawn?

1 $y = -x^2, y = -0.5x^2, y = -2x^2$

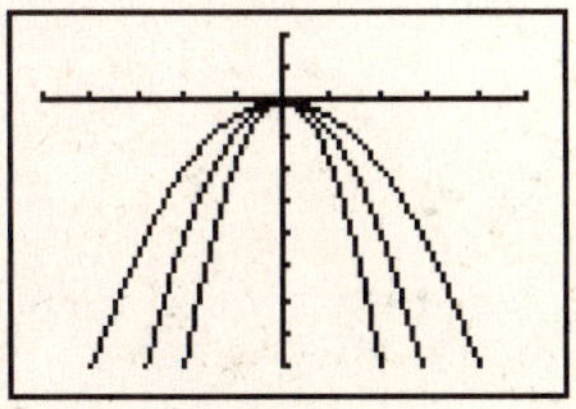

Each graph opens ________ and has its vertex at the ________.

The graph of $y = -0.5x^2$ is ________ than the graph of $y = -x^2$.

The graph of $y = -2x^2$ is ________ than the graph of $y = -x^2$.

The shape of the parabola narrows as the coefficient of $-x^2$ becomes greater. The shape widens as the coefficient of $-x^2$ becomes smaller.

2 $y = -x^2, y = -x^2 + 1, y = -x^2 + 4$

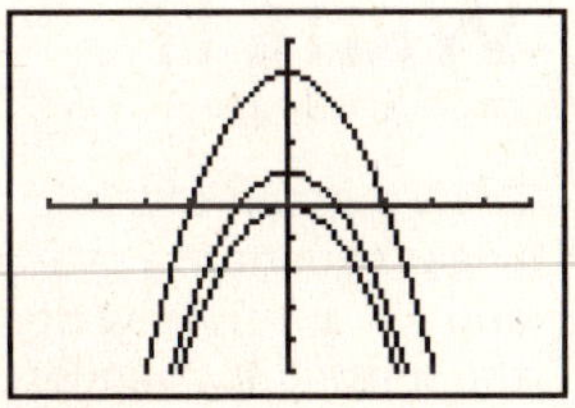

Each graph opens ________ and has the same shape as $y = -x^2$ so they form a ________.

Each parabola has a different ________ located along the y-axis.

A constant greater than 0 shifts the graph ________ along the axis of ________.

3 $y = x^2, y = (x - 3)^2, y = (x + 1)^2$

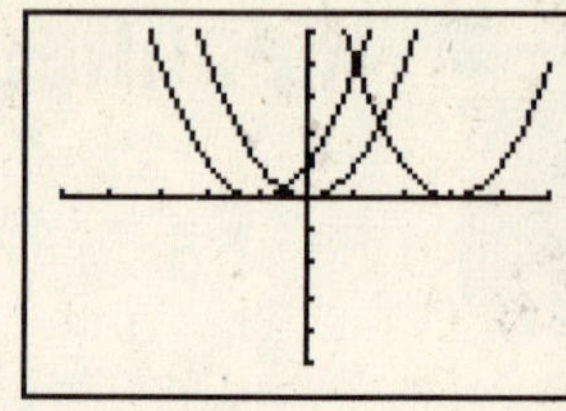

Each graph opens ____ and has the same shape as $y = x^2$.

However, each parabola has a different ____ located along the ____.

Find the number for x that results in ____ inside the parentheses. The graph shifts this number of ____ to the ____ or ____.

4 $y = x^2, y = (x + 2)^2 - 1$

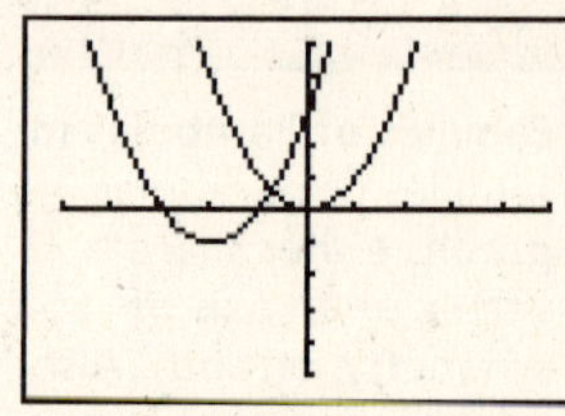

The graph of $y = (x + 2)^2 - 1$ has the same shape as the graph of ____.

However, it shifts to the ____ two units because a negative ____ will result in ____ inside the parentheses. It also shifts ____ 1 unit because of the constant ____ outside the parentheses.

WRITE IT

Draw a graph of the linear equations $y = x$ and $y = x + 1$. How do the shifting techniques learned in this lesson relate to your observations of these two linear graphs?

______

______

______

______

______

______

Your Turn **Each group of equations is graphed on the same coordinate plane. Compare and contrast the graphs. What conclusions can be drawn?**

a. $y = x^2, y = 0.5x^2, y = 3x^2$

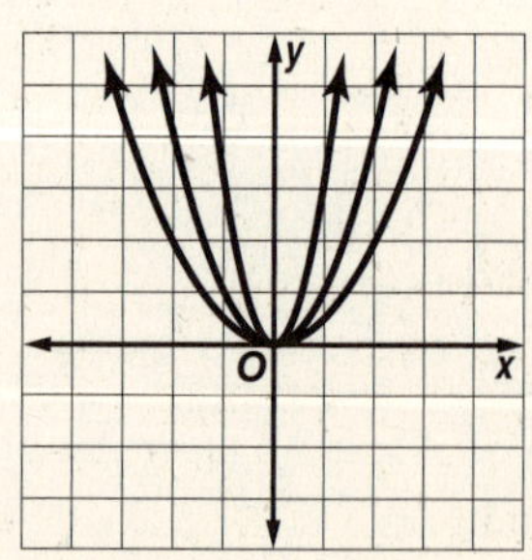

b. $y = 2x^2, y = 2x^2 - 3, y = 2x^2 + 2.$

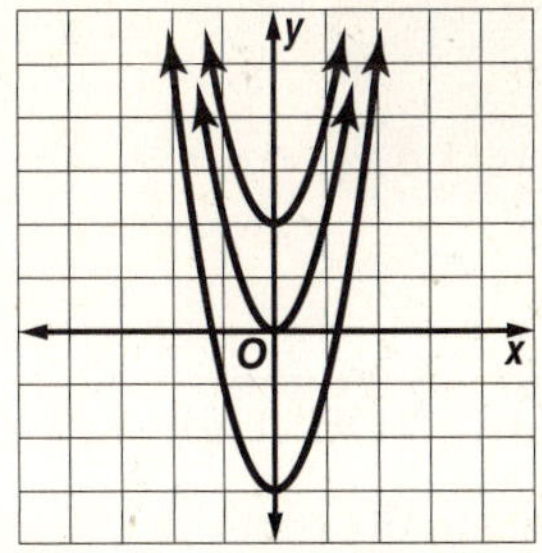

c. $y = -x^2, y = -(x + 2)^2, y = -(x - 3)^2$

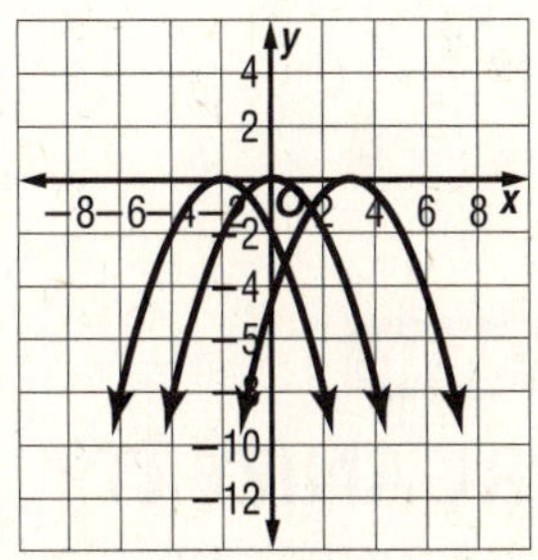

d. $y = x^2, y = (x + 1)^2 + 2$

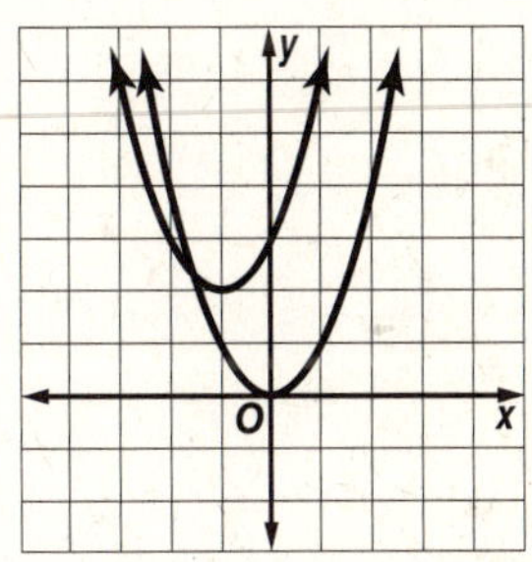

HOMEWORK ASSIGNMENT

Page(s):

Exercises:

11-3 Solving Quadratic Equations by Graphing

What You'll Learn

- Locate the roots of quadratic equations by graphing the related functions.

Build Your Vocabulary (page 223)

A **quadratic equation** is an equation of the form

______, where $a \neq$ ______.

The ______ of a quadratic equation are called the **roots** of the equation.

The ______ of a quadratic function are called **zeros**.

Foldables™

Organize It

Draw a parabola and label the zeros under the tab for Lesson 11-3.

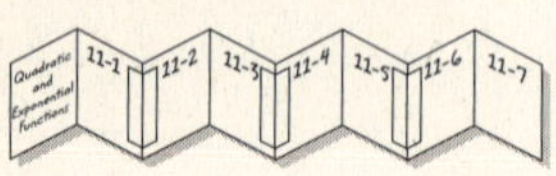

Example

1 Find the roots of $x^2 + 2x - 15 = 0$ by graphing the related function.

Graph the ______ function $f(x) =$ ______.

Before making a table of values, find the equation of the ______ of symmetry. This will make selecting ______ for your table easier.

$x = -\frac{b}{2a}$ Equation of the axis of symmetry

$x = -\frac{\boxed{}}{2(\boxed{})}$ or ______ $a =$ ______, $b =$ ______

The equation of the axis of symmetry is ______. Now, make a table using x-values around ______. Graph each point on a coordinate plane.

x	$x^2 + 2x - 15$	$f(x)$
-5	$(-5)^2 + 2(-5) - 15$	
-3	$(-3)^2 + 2(-3) - 15$	
-1	$(-1)^2 + 2(-1) - 15$	
1	$(1)^2 + 2(1) - 15$	
3	$(3)^2 + 2(3) - 15$	

The zeros of the function appear to be ____ and ____.

So, the roots are ____ and ____.

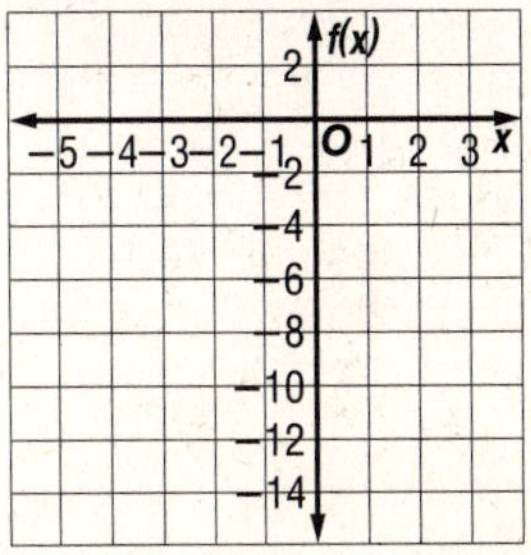

Your Turn Find the roots of $-x^2 + 4x + 12 = 0$ by graphing the related function.

EXAMPLE

2 **Estimate the roots of $-x^2 + 4x - 1 = 0$.**

Find the equation of the axis of symmetry.

$x = -\frac{b}{2a}$ Equation of the axis of symmetry

$x = -\frac{\square}{2(\square)}$ or ____ $a =$ ____, $b =$ ____

The equation of the axis of symmetry is ____. Now, make a table using x-values around ____. Graph each point on a coordinate plane.

x	$-x^2 + 4x - 1$	$f(x)$
0	$-0^2 + 4(0) - 1$	
1	$-1^2 + 4(1) - 1$	
2	$-2^2 + 4(2) - 1$	
3	$-3^2 + 4(3) - 1$	2
4	$-4^2 + 4(4) - 1$	-1

The x-intercepts of the graph are between 0 and 1 and between 3 and 4. So, one root of the equation is between ______ and the other root is between ______.

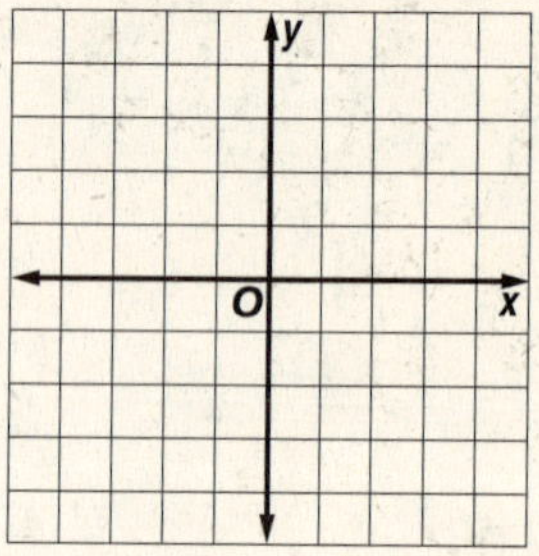

Your Turn Estimate the roots of $x^2 + 3x - 2 = 0$.

______

HOMEWORK ASSIGNMENT

Page(s):

Exercises:

11–4 Solving Quadratic Equations by Factoring

What You'll Learn

- Solve quadratic equations by factoring and by using the Zero Product Property.

Key Concept

Zero Product Property
For all numbers a and b, if $ab = 0$, then $a = 0$, $b = 0$, or both a and b equal 0.

Foldables Summarize the Zero Product Property in your own words under the tab for Lesson 11-4.

EXAMPLE

1 Solve $-2x(x + 5) = 0$. Check your solution.

If $-2x(x + 5) = 0$, then ______ $= 0$ or ______ $= 0$.

______ $= 0$ or ______ $= 0$

$x =$ ______ or $x =$ ______

Check: Substitute each value for x in the original equation.

$-2x(x + 5) = 0$ $\qquad$ $-2x(x + 5) = 0$

$-2(\_\_)(\_\_ + 5) \stackrel{?}{=} 0$ $\qquad$ $-2(\_\_)(\_\_ + 5) \stackrel{?}{=} 0$

$0(\_\_) \stackrel{?}{=} 0$ $\qquad$ $\_\_(0) \stackrel{?}{=} 0$

$\_\_ = 0$ $\qquad$ $\_\_ = 0$

Your Turn Solve $-3x(x - 4) = 0$. Check your solution.

______

EXAMPLE

2 A child throws a ball up in the air. The height h of the ball t seconds after it has been thrown is given by the equation $h = -16t^2 + 8t + 4$. Solve $4 = -16t^2 + 8t + 4$ or $8t(-2t + 1) = 0$ to find how long it would take the ball to reach the height from which it was thrown.

If $8t(-2t + 1) = 0$, then ______ $= 0$ or ______ $= 0$.

______ $= 0$ or ______ $= 0$

$t =$ ______ $\qquad$ $-2t =$ ______

$t =$ ______

The solutions are ☐ and ☐. The solution 0 represents the beginning of the throw. So, the ball would reach the height from which it was thrown after ☐ of a second.

Your Turn A child throws a ball up in the air. The height h of the ball t seconds after it has been thrown is given by the equation $h = -16t^2 + 4t + 3$. Solve $3 = -16t^2 + 4t + 3$ or $4t(-4t + 1) = 0$ to find how long it would take the ball to reach the height from which it was thrown.

EXAMPLE

3 Solve $x^2 - 4x - 21 = 0$. Check your solution.

$x^2 - 4x - 21 = 0$		
$(x + \square)(x - \square) = 0$		Factor
$\square = 0$ or $\square = 0$		Zero Product Property
$x = \square$ or $x = \square$		

Check:

$x^2 - 4x - 21 = 0$	$x^2 - 4x - 21 = 0$
$(\square)^2 - 4(\square) - 21 = 0$	$\square^2 - 4(\square) - 21 = 0$
$\square + \square - 21 = 0$	$\square - \square - 21 = 0$
$\square = 0$	$\square = 0$

Your Turn Solve $x^2 - 11x + 30 = 0$. Check your solution.

HOMEWORK ASSIGNMENT

Page(s):

Exercises:

11–5 Solving Quadratic Equations by Completing the Square

What You'll Learn

- Solve quadratic equations by completing the square

Build Your Vocabulary (page 222)

A method to make any ______ expression a ______ square is called **completing the square**.

Foldables™ Organize It

Create your own quadratic equation (choose any integers for *a*, *b*, and *c*) and write it under the tab for Lesson 11-5. Solve the equation by completing the square.

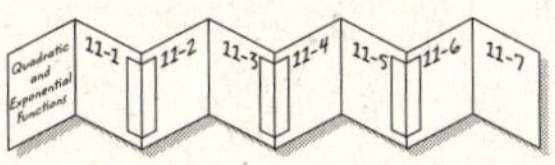

Example

1 Find the value of c that makes $x^2 - 8x + c$ a perfect square.

Step 1 Find half of -8.

$$\frac{\boxed{}}{2} = -4$$

Step 2 Square the result of Step 1.

$$(-4)^2 = \boxed{}$$

Step 3 Add the result of Step 2 to $x^2 - 8x$.

$$x^2 - 8x + \boxed{}$$

Thus, $c = \boxed{}$.

Notice that $x^2 - 8x + 16 = \left(\boxed{}\right)^2$.

Your Turn Find the value of c that makes $x^2 + 12x + c$ a perfect square.

Remember It

When the square root is taken on both sides of an equation, it is necessary to add a ± sign.

EXAMPLE

2 Solve $x^2 + 4x - 5 = 0$ by completing the square.

$x^2 + 4x - 5 = 0$	$x^2 + 4x - 5$ is not a perfect square.
$x^2 + 4x = \square$	Add $\square$ to each side.
$x^2 + 4x + \square = 5 + \square$	Since $\left(\frac{\square}{2}\right)^2 = \square$, add $\square$ to each side.
$\left(x + \square\right)^2 = \square$	Factor $x^2 + 4x + 4$.
$\sqrt{\left(x + \square\right)^2} = \pm\sqrt{\square}$	Take the square root of each side.
$x + \square = \pm\square$	
$x + 2 - \square = \pm 3 - \square$	Subtract $\square$ from each side.
$x = \square$ or $x = \square$	Simplify each equation.
$x = \square$ or $x = \square$	Check the solution.

The solutions are $\square$ and $\square$.

Your Turn Solve $x^2 - 4x - 21 = 0$ by completing the square.

HOMEWORK ASSIGNMENT

Page(s):

Exercises:

11–6 The Quadratic Formula

What You'll Learn

- Solve quadratic equations by using the Quadratic Formula.

Key Concept

The Quadratic Formula

For $ax^2 + bx + c = 0$,

$x = \dfrac{-b \pm \sqrt{b^2 - 4ac}}{2a}$,

$a \neq 0$.

Foldables Write the Quadratic Formula under the tab for Lesson 11-6. Use the formula to solve the quadratic equation that you created under the tab for Lesson 11-5.

EXAMPLES

Use the Quadratic Formula to solve each equation.

1 $2x^2 - 5x + 3 = 0$

$x = \dfrac{-b \pm \sqrt{b^2 - 4ac}}{2a}$, $a =$ ___, $b =$ ___, and $c =$ ___

$$x = \frac{-(\_\_\_) \pm \sqrt{(\_\_\_)^2 - 4(\_\_\_)(\_\_\_)}}{2(\_\_\_)}$$

$$= \frac{\_\_\_ \pm \sqrt{\_\_\_ - \_\_\_}}{\_\_\_} = \frac{5 \pm \sqrt{\_\_\_}}{4} \text{ or } \frac{5 \pm \_\_\_}{4}$$

$$x = \frac{5 + \_\_\_}{4} \quad \text{or} \quad x = \frac{5 - \_\_\_}{4}$$

$$x = \frac{\_\_\_}{4} \text{ or } \_\_\_ \qquad x = \frac{\_\_\_}{4} \text{ or } \_\_\_$$

2 $-x^2 + 6x - 9 = 0$

$x = \dfrac{-b \pm \sqrt{b^2 - 4ac}}{2a}$, $a =$ ___, $b =$ ___, and $c =$ ___

$$= \frac{-(\_\_\_) \pm \sqrt{(\_\_\_)^2 - 4(\_\_\_)(\_\_\_)}}{2(\_\_\_)}$$

$$= \frac{\_\_\_ \pm \sqrt{\_\_\_ - \_\_\_}}{\_\_\_}$$

$$= \frac{-6 \pm \_\_\_}{-2} = \frac{\_\_\_}{-2} \text{ or } \_\_\_$$

Your Turn **Use the Quadratic Formula to solve each equation.**

a. $2x^2 - 3x - 2 = 0$

b. $6n^2 - 7n - 3 = 0$

EXAMPLE

3 **"Hang time" is the total amount of time a ball stays in the air. A punter kicks the football with an upward velocity of 58 ft/s and his foot meets the ball 1 foot off the ground. His formula is $h(t) = -16t^2 + 58t + 1$, where $h(t)$ is the ball's height for any time t after the ball was kicked. What is the hang time?**

$$t = \frac{-b \pm \sqrt{b^2 - 4ac}}{2a},\ a = [\ ],\ b = [\ ],\ \text{and } c = [\ ]$$

$$= \frac{-[\ ] \pm \sqrt{([\ ])^2 - 4([\ ])(1)}}{2(-16)}$$

$$= \frac{-58 \pm \sqrt{[\ ] + [\ ]}}{[\ ]}$$

$$t = \frac{-58 + [\ ]}{-32} \quad \text{or} \quad t = \frac{-58 - [\ ]}{-32}$$

$$\approx [\ ] \quad \text{or} \quad \approx [\ ]$$

Since time cannot be negative, the only approximate solution is [ ]. The football has a hang time of about [ ].

REMEMBER IT

The square root of a negative number is not a real number. This means that there is no real solution for x.

HOMEWORK ASSIGNMENT

Page(s):

Exercises:

Your Turn A baseball player hits a baseball with an upward velocity of 50 feet per second from a height of 3 feet. The equation $y = -16t^2 + 50t + 3$ gives the height y of the ball after t seconds. How long will it take the ball to be 5 feet above the ground on the way down?

11-7 Exponential Functions

WHAT YOU'LL LEARN

- Graph exponential functions.

BUILD YOUR VOCABULARY (page 222)

A function in which the ______ is a variable is an **exponential function**.

The **initial value** of an exponential function is the value of the function when ______.

FOLDABLES™

ORGANIZE IT

Sketch an exponential function under the tab for Lesson 11-7. Label the initial value.

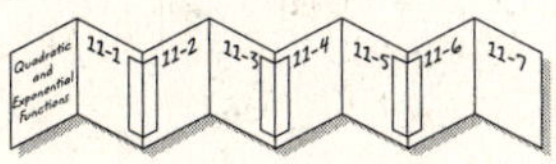

EXAMPLE

1 Graph $y = 1.5^x$.

x	1.5^x	y
-1	1.5^{-1}	
0	1.5^0	
1	1.5^1	
2	1.5^2	
3	1.5^3	

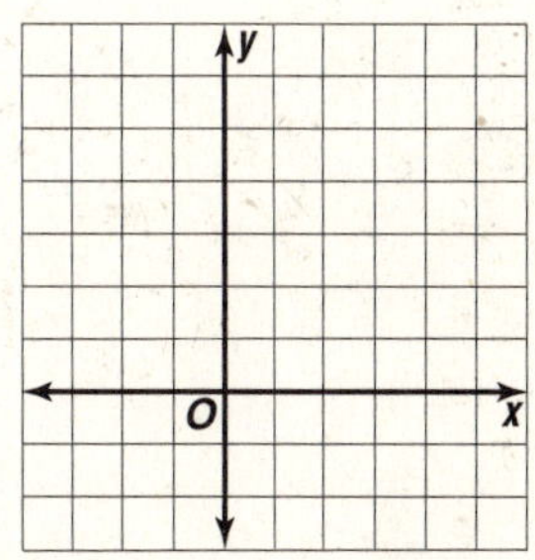

Your Turn Graph $y = 3^x$.

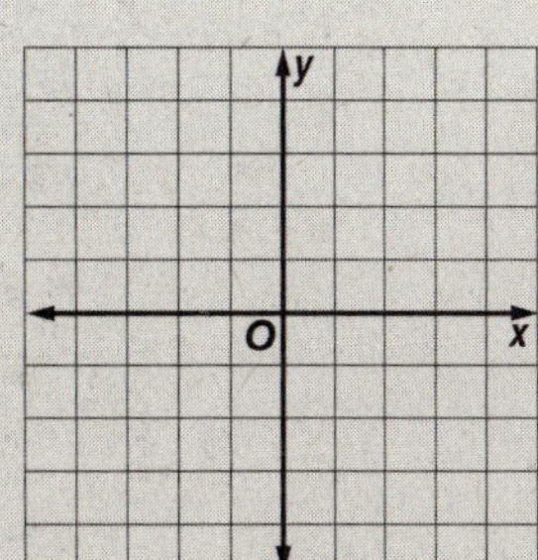

EXAMPLE

2 **Graph the exponential function. Then state the y-intercept.**

$y = 3^x + 1$

x	$3^x + 1$	y
-1	$3^{-1} + 1$	
0	$3^0 + 1$	
1	$3^1 + 1$	
2	$3^2 + 1$	
3	$3^3 + 1$	

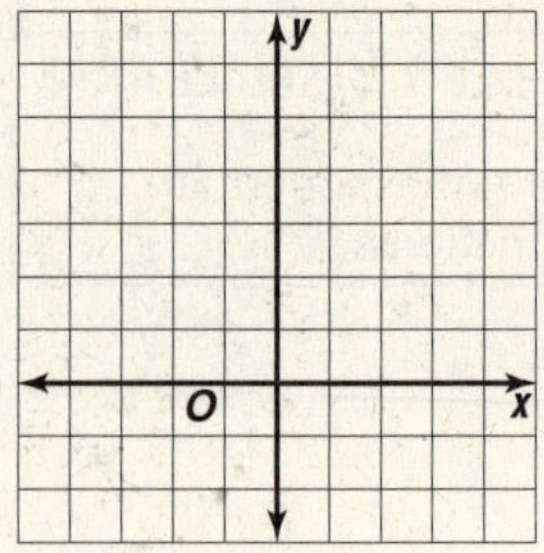

To find the y-intercept, let $x =$ ☐ and solve for y.

$y = 3^0 + 1$ or ☐

So, the y-intercept is ☐.

Your Turn **Graph each exponential function. Then state the y-intercept.**

a. $y = 2^x - 1$

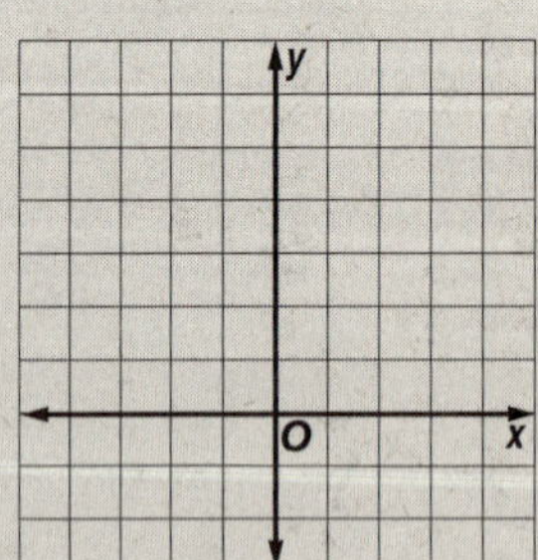

b. $y = 3^x + 2$

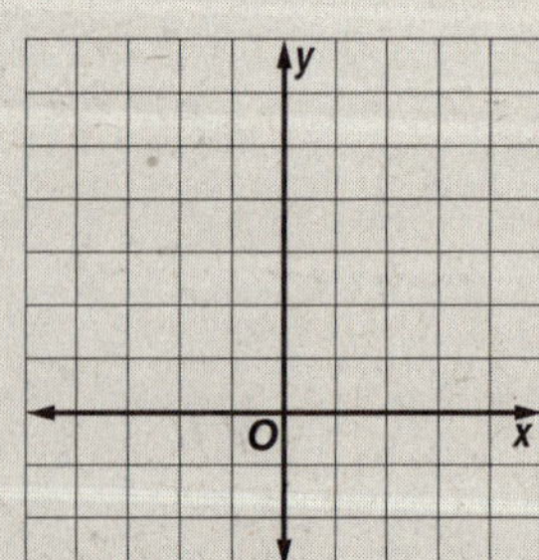

HOMEWORK ASSIGNMENT

Page(s):

Exercises:

CHAPTER 11

BRINGING IT ALL TOGETHER

STUDY GUIDE

FOLDABLES	VOCABULARY PUZZLEMAKER	BUILD YOUR VOCABULARY
Use your **Chapter 11 Foldable** to help you study for your chapter test.	To make a crossword puzzle, word search, or jumble puzzle of the vocabulary words in Chapter 11, go to: www.glencoe.com/sec/math/t_resources/free/index.php	You can use your completed **Vocabulary Builder** (pages 222–223) to help you solve the puzzle.

11-1 Graphing Quadratic Functions

Complete each statement about the graph at the right.

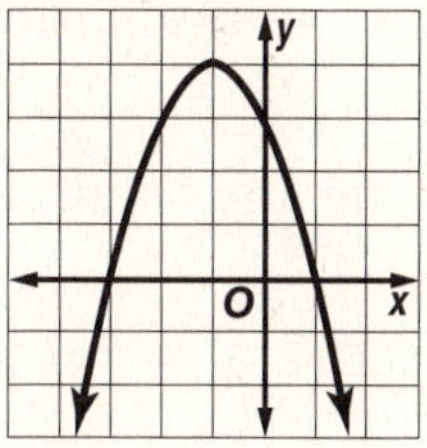

1. The graph is a curve called a ________.

2. The highest point of the graph is located at ________.

3. This point is the ________ (maximum/minimum) point of the graph.

4. If you fold a parabola along a line to get two halves that match exactly, the line where you fold the parabola is the ________ of the parabola. This line goes through the ________ of the parabola.

11-2 Families of Quadratic Functions

Refer to these parabolas that were graphed on a calculator.

5. Do the parabolas graphed form a family of parabolas? Explain why or why not.

____

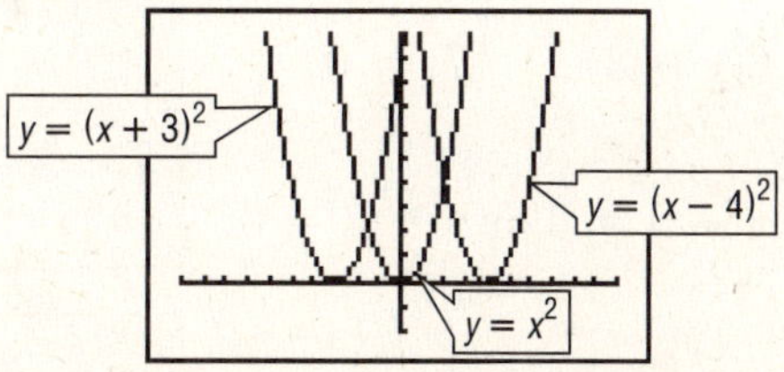

11-3 Solving Quadratic Equations by Graphing

The graphs of two functions are shown. Use the graphs to provide the requested information about the related quadratic equations.

a. $f(x) = x^2 + 2x + 3$

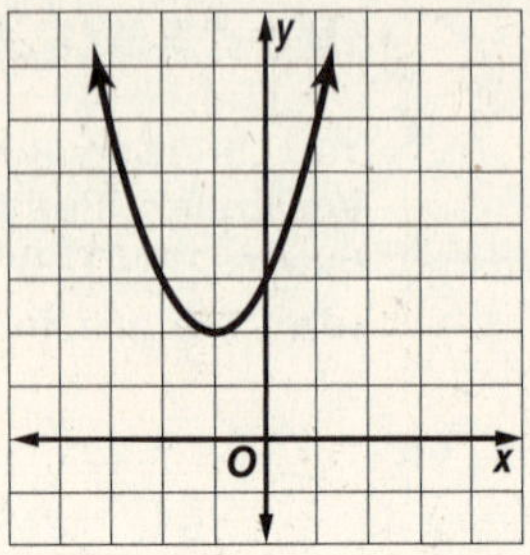

b. $f(x) = x^2 + x - 2$

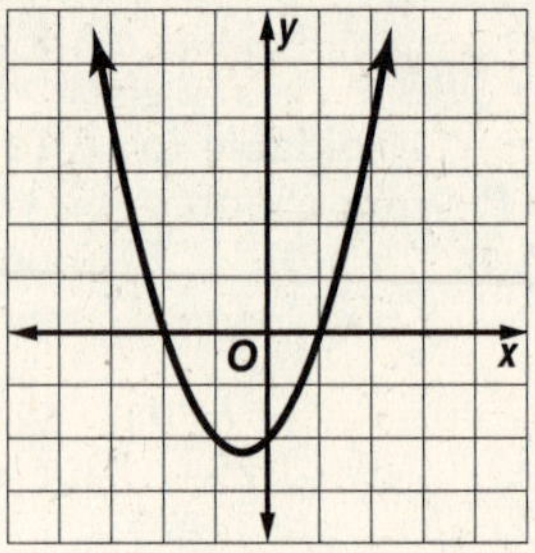

6. For Graph a, the related quadratic equation is ______.

7. How many solutions are there? ______

8. Name any solutions. ______

9. For Graph b, the related quadratic equation is ______.

10. How many solutions are there? ______

11. Name any solutions. ______

11-4 Solving Quadratic Equations by Factoring

Provide a reason for each step in the solution of the equation.

$\mathbf{2x^2 + 8x = 42}$

12. $2x^2 + 8x - 42 = 0$ ______

13. $2(x^2 + 4x - 21) = 0$ ______

14. $2(x - 3)(x + 7) = 0$ ______

15. $x - 3 = 0$ or $x + 7 = 0$ ______

16. $x = 3$ or $x = -7$ ______

11-5 Solving Quadratic Equations by Completing the Square

Find the value of c that makes each trinomial a square.

17. $x^2 + 2x + c$

18. $x^2 - 9x + c$

19. Solve $x^2 - 6x - 72 = 0$ by completing the square.

11-6 The Quadratic Formula

20. Solve $12x^2 + 7x = 15$ using the Quadratic Formula. Round to the nearest hundredth.

21. Find the length and width of the rectangle shown.

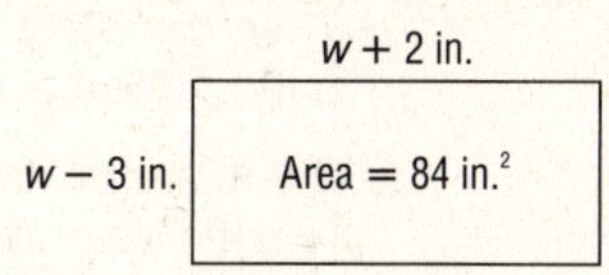

11-7 Exponential Functions

Match the correct equation with each graph shown.

22.

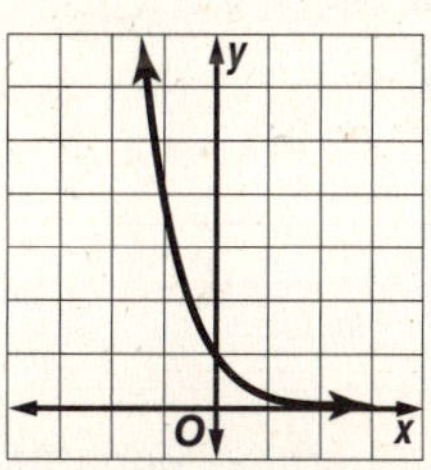

23.

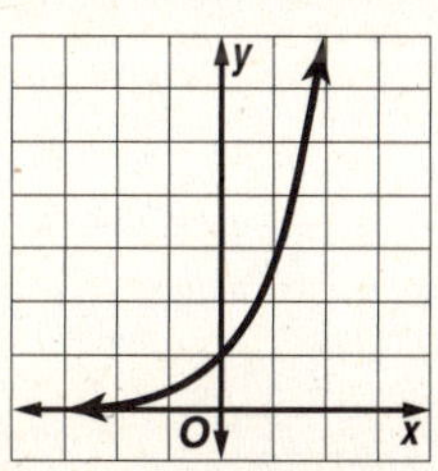

24.

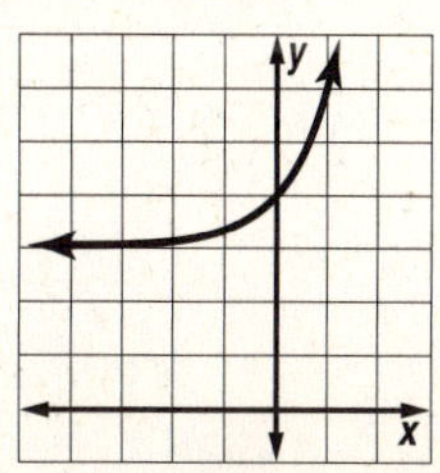

a. $y = 3^x$ **b.** $y = 4^{-x}$ **c.** $y = 3^x + 3$ **d.** $y = -4^x$

CHAPTER 11

Checklist

ARE YOU READY FOR THE CHAPTER TEST?

Math Online

Visit **algconcepts.com** to access your textbook, more examples, self-check quizzes, and practice tests to help you study the concepts in Chapter 11.

Check the one that applies. Suggestions to help you study are given with each item.

☐ **I completed the review of all or most lessons without using my notes or asking for help.**

- You are probably ready for the Chapter Test.
- You may want take the Chapter 11 Practice Test on page 499 of your textbook as a final check.

☐ **I used my Foldable or Study Notebook to complete the review of all or most lessons.**

- You should complete the Chapter 11 Study Guide and Review on pages 496–498 of your textbook.
- If you are unsure of any concepts or skills, refer back to the specific lesson(s).
- You may also want to take the Chapter 11 Practice Test on page 499.

☐ **I asked for help from someone else to complete the review of all or most lessons.**

- You should review the examples and concepts in your Study Notebook and Chapter 11 Foldable.
- Then complete the Chapter 11 Study Guide and Review on pages 496–498 of your textbook.
- If you are unsure of any concepts or skills, refer back to the specific lesson(s).
- You may also want to take the Chapter 11 Practice Test on page 499.

Student Signature

Parent/Guardian Signature

Teacher Signature

CHAPTER 12

Inequalities

FOLDABLES™ Use the instructions below to make a Foldable to help you organize your notes as you study the chapter. You will see Foldable reminders in the margin of this Interactive Study Notebook to help you in taking notes.

Begin with four sheets of grid paper.

STEP 1 **Fold**
Fold each sheet in half from top to bottom.

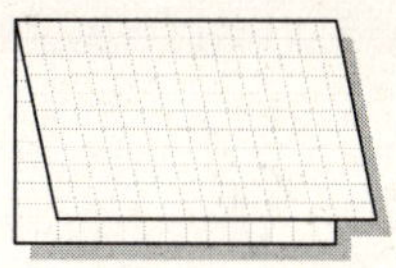

STEP 2 **Cut**
Cut along fold. Staple the eight half-sheets together to form a booklet.

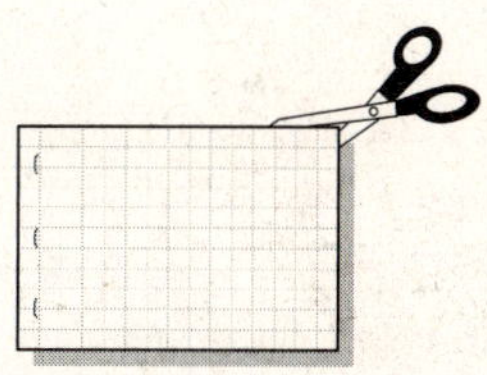

STEP 3 **Label**
Label each page with a lesson number and title.

NOTE-TAKING TIP: When you take notes, define new terms and write about the new concepts in your own words. Write your own examples that use the new terms and concepts.

CHAPTER 12

BUILD YOUR VOCABULARY

This is an alphabetical list of new vocabulary terms you will learn in Chapter 12. As you complete the study notes for the chapter, you will see Build Your Vocabulary reminders to complete each term's definition or description on these pages. Remember to add the textbook page number in the second column for reference when you study.

Vocabulary Term	Found on Page	Definition	Description or Example
boundary			
compound inequality			
half-plane			
intersection			
quadratic inequalities			
set-builder notation			
union			

12–1 Inequalities and Their Graphs

What You'll Learn

- Graph inequalities on a number line.

EXAMPLE

1 **Many movie theaters give a senior-citizen discount to people who are 65 or over. Write an inequality that describes those who are eligible to receive the discount.**

Let a represent the ages of people who are eligible to receive the discount.

The ages of all those eligible	*are greater than or equal*	*to 65 years.*
_____	_____	_____

_____ is the same as $65 \leq a$.

Then a is _____ 65.

Your Turn In Colorado, the speed limit on an interstate highway is 75 miles per hour. Write an inequality that describes the speed cars are allowed to travel.

_____

EXAMPLES

Graph each inequality on a number line.

Remember It

A bullet, or closed circle, is used when the inequality includes the endpoint. An open circle is used when the inequality does not include the endpoint.

2 $x \leq -1$

Since x can equal _____, graph a _____ at _____ and shade to the _____.

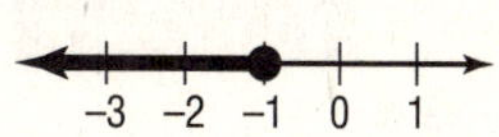

3 $k > \frac{3}{4}$

Since x cannot equal _____, graph a _____ at _____ and shade to the _____.

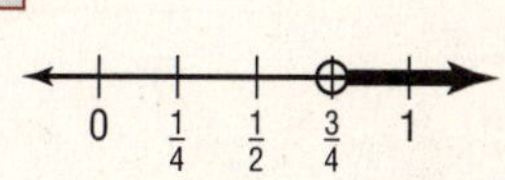

Your Turn **Graph each inequality on a number line.**

a. $b > -1.8$

b. $w \le 5$

EXAMPLES

FOLDABLES™

ORGANIZE IT

Summarize the meaning of the following signs: <, ≤, >, and ≥ under the tab for Lesson 12–1.

4 Write an inequality for the graph.

Locate where the graph begins. This graph begins at ____, and 2 is included. Also note that the arrow points to the ____. The graph describes values that are ____ 2. So, ____.

5 Write an inequality for the graph.

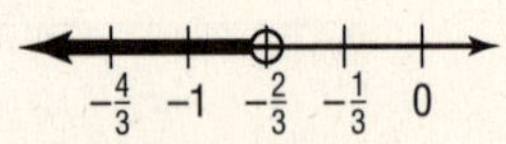

Locate where the graph begins. This graph begins at ____, and $\frac{2}{3}$ is not included. Note that the arrow points to the ____. The graph describes values that are ____. So, ____.

Your Turn **Write an inequality for each graph.**

a.

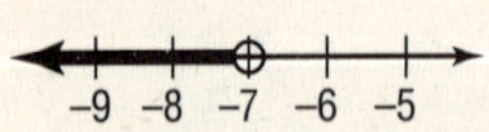

b.

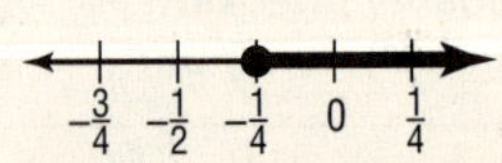

EXAMPLE

6 No more than 12 students can fit in a large school van. Write an inequality to express this information. Then graph the inequality on a number line.

Let s represent the ____ who can fit in a large school van. Then write an inequality using the ____ symbol since the number ____ 12 students. ____

To graph the inequality, first graph a ____ at ____. Then include the numbers ____ than ____ by drawing and shading an ____ to the ____.

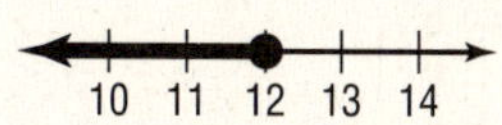

Your Turn A classroom can seat no more than 30 people. Write an inequality to represent this situation. Then graph the inequality on a number line.

____

HOMEWORK ASSIGNMENT

Page(s):

Exercises:

12–2 Solving Addition and Subtraction Inequalities

EXAMPLE

WHAT YOU'LL LEARN

- Solve inequalities involving addition and subtraction.

KEY CONCEPT

Addition and Subtraction Properties for Inequalities For any inequality, if the same quantity is added or subtracted to each side, the resulting inequality is true.

1 Solve $x - 5 < -2$. Check your solution.

$$x - 5 < -2$$

$x - 5 +$ ______ $< -2 +$ ______ Add ______ to each side.

$x <$ ______

Check: Substitute a number less than ______, the number ______, and a number greater than ______ into the inequality.

Let $x = 0$.	Let $x = 3$.	Let $x = 5$.
$y - 5 < -2$	$y - 5 < -2$	$y - 5 < -2$
______ $- 5 \stackrel{?}{<} -2$	______ $- 5 \stackrel{?}{<} -2$	______ $- 5 \stackrel{?}{<} -2$
______ $\stackrel{?}{<} -2$	______ $\stackrel{?}{<} -2$	______ $\stackrel{?}{<} -2$
______	______	______

The solution is {all numbers ______ }.

Your Turn Solve $r - 12 \leq -2$. Check your solution.

______

BUILD YOUR VOCABULARY (page 246)

Set-builder notation is a method of writing the solution set for an ______; $\{x \mid x > 3\}$.

EXAMPLE

2 **Suppose that you plan to run at least 7.5 miles per week during the summer to train for the cross-country season in the fall. During one week, you run 2 miles on Sunday and 2.5 miles on Wednesday. Solve $2 + 2.5 + m \geq 7.5$ to find out how many more miles need to be run before the week is over.**

$2 + 2.5 + m \geq 7.5$

____ $+\ m \geq 7.5$

$4.5 + m -$ ____ $\geq 7.5 -$ ____ Subtract ____ from each side.

$m \geq$ ____

The solution can be written as $\{m \mid$ ____ $\}$. So, at least ____ more miles need to be run before the week is over.

Your Turn Suppose that your teacher recommends that you read 3 hours each week. During one week, you read 35 minutes on Monday and 25 minutes on Tuesday. Solve $35 + 25 + m \geq 180$ to find how many more minutes m you must read this week for your weekly total to be at least 3 hours.

FOLDABLES™

ORGANIZE IT

Write the solution set for Example 3 using words under the tab for Lesson 12-2.

REMEMBER IT

The vertical line, |, used in set builder notation can be read "such that".

HOMEWORK ASSIGNMENT

Page(s):

Exercises:

12–3 Solving Multiplication and Division Inequalities

What You'll Learn

- Solve inequalities involving multiplication and division.

Key Concept

Division Property for Inequalities If you divide each side of an inequality by a positive number, the inequality remains true. If you divide each side of an inequality by a negative number, the inequality symbol must by reversed for the inequality to remain true.

EXAMPLE

1 An electric car that needs to be recharged every 260 miles should travel no more than 130 miles from the charger. If you drive at an average speed of 50 miles per hour, what are the lengths of time t you can drive away from the charger and then still make it back without running out of energy?

Recall that $rt = d$.

$\_\_\_ t \leq \_\_\_$

$\frac{50t}{\_\_\_} \leq \frac{130}{\_\_\_}$ Divide each side by ___.

$t \leq \_\_\_$ The inequality symbol is facing the same direction.

You should travel no more than ___ hours or 2 hours and ___ minutes away from the charger.

Your Turn Two friends walk at least 3 miles every day. If they walk 2 miles per hour, how long do they walk every day?

___

EXAMPLE

2 Solve $-2.5x > 10$. Check your solution.

$-2.5x > 10$

$\frac{-2.5x}{\_\_\_} < \frac{10}{\_\_\_}$ Divide each side by ___ and reverse the inequality symbol.

$x < \_\_\_$

Check: Substitute ______ and a number less than ______, such as −5, into the original inequality.

Let $x = -4$.

$$-2.5x > 10$$

$$-2.5(\ \_\_\_\_\ ) \overset{?}{>} 10$$

$$\_\_\_\_ > 10 \quad \_\_\_\_$$

Let $x = -5$.

$$-2.5x > 10$$

$$-2.5(\ \_\_\_\_\ ) \overset{?}{>} 10$$

$$\_\_\_\_ > 10 \quad \_\_\_\_$$

The solution set is $\{x \mid$ ______ $\}$.

EXAMPLE

3 Solve $-\frac{1}{2}x \geq -5$. Check your solution.

$$-\frac{1}{2}x \geq -5$$

$$\_\_\_\_\left(-\frac{1}{2}x\right) \leq \_\_\_\_(-5)$$ Multiply each side by −2 and reverse the inequality symbol.

$$x \leq \_\_\_\_$$

Check: Substitute ______ and a number less than ______, such as 8, into the original inequality.

Let $x = 10$.

$$-\frac{1}{2}x \geq -5$$

$$-\frac{1}{2}(\ \_\_\_\_\ ) \overset{?}{\geq} -5$$

$$\_\_\_\_ \geq -5 \quad \_\_\_\_$$

Let $x = 8$.

$$-\frac{1}{2}x \geq -5$$

$$-\frac{1}{2}(\ \_\_\_\_\ ) \overset{?}{\geq} -5$$

$$\_\_\_\_ \geq -5 \quad \_\_\_\_$$

The solution is $\{x \mid$ ______ $\}$.

KEY CONCEPT

Multiplication Property for Inequalities If you multiply each side of an inequality by a positive number, the inequality remains true. If you multiply each side of an inequality by a negative number, the inequality symbol must be reversed for the inequality to remain true.

FOLDABLES Write the Multiplication and Division Properties for Inequalities under the tab for Lesson 12–3.

Your Turn **Solve each inequality. Check your solution.**

a. $-3r < 21$

b. $-\frac{3}{4}y > 12$

HOMEWORK ASSIGNMENT

Page(s):

Exercises:

12–4 Solving Multi-Step Inequalities

WHAT YOU'LL LEARN

- Solve inequalities involving more than one operation.

EXAMPLE

1 Solve $5x - 9 > 21$. Check your solution.

$5x - 9 > 21$

$5x - 9 +$ ______ $> 21 +$ ______ Add ______ to each side.

$5x >$ ______

$\frac{5x}{5} > \frac{30}{5}$ Divide each side by 5.

$x >$ ______

Check: Substitute 6 and 7 into the original inequality.

$5x - 9 > 21$	$5x - 9 > 21$
$5(6) - 9 \overset{?}{>} 21$	$5(7) - 9 \overset{?}{>} 21$
$30 - 9 \overset{?}{>} 21$	$35 - 9 \overset{?}{>} 21$
______ > 21 false	______ > 21 true

The solution is {x | ______ }.

EXAMPLE

2 Solve $16 - 2x \leq 3x + 1$. Check your solution.

$16 - 2x \leq 3x + 1$

$16 - 2x -$ ______ $\leq 3x + 1 -$ ______

$16 - 5x \leq 1$

$16 - 5x -$ ______ $\leq 1 -$ ______

$-5x \leq -15$

$\frac{-5x}{-5} \geq \frac{-15}{-5}$ Divide and reverse the inequality symbol.

$x \geq$ ______

FOLDABLES™

ORGANIZE IT

Write and solve an inequality requiring more than one step to solve under the tab for Lesson 12–4. Then, name the operations used.

Check: Substitute 3 and 4 into the original inequality.

Let $x = 3$.	Let $x = 4$.
$16 - 2x \le 3x + 1$	$16 - 2x \le 3x + 1$
$16 - 2(3) \stackrel{?}{\le} 3(3) + 1$	$16 - 2(4) \stackrel{?}{\le} 3(4) + 1$
$10 \le \square$ true	$8 \le \square$ true

The solution is $\{x \mid \square\}$.

EXAMPLE

3 Solve $3(x + 2) > -75$. Check your solution.

$3(x + 2) > -75$

$3x + \square > -75$ Distributive Property

$3x + 6 - \square > -75 - \square$ Subtract $\square$ from each side.

$3x > \square$

$\frac{3x}{\square} > \frac{-81}{\square}$ Divide each side by $\square$.

$x > \square$

The solution is $\{x \mid \square\}$. Check your solution.

Your Turn **Solve each inequality. Check your solution.**

a. $-3t + 2 > -4$

b. $10 + 3x > 5x - 4$

c. $15 > -5(x + 6)$

EXAMPLE

4 **Karl's point totals in the first four of five basketball games were 15, 12, 19, and 18. How many points *t* must he score in the fifth game to have a mean point total of more than 16?**

The sum of Karl's points, divided by ______, will give the mean score. The mean must be more than ______.

$$\frac{15 + 12 + 19 + 18 + t}{\square} > \square$$

$$\square\left(\frac{15 + 12 + 19 + 18 + t}{\square}\right) > \square(16)$$ Multiply each side by ______.

$$15 + 12 + 19 + 18 + t > \square$$

$$\square + t > 80$$

$$64 + t - \square > 80 - \square$$ Subtract.

$$t > \square$$

Karl must score more than ______ points in the fifth game to have a mean point total of more than ______.

Your Turn Lien's score on the first four of five 100-point tests were 82, 85, 95, and 91. What score *s* on the fifth test will give her a mean score of at least 90 for all five tests?

HOMEWORK ASSIGNMENT

Page(s):

Exercises:

12–5 Solving Compound Inequalities

WHAT YOU'LL LEARN

- Solve compound inequalities.

BUILD YOUR VOCABULARY (page 246)

Two or more inequalities that are connected by the words ____ or ____ form a **compound inequality.**

An **intersection** is the set of elements common to ____ inequalities.

A **union** is the set of elements in each of ____ inequalities.

FOLDABLES™

ORGANIZE IT

Summarize the difference between "intersection" and "union" under the tab for Lesson 12–5. Then, give examples of when to use each one.

EXAMPLE

1 **Write $x < 0$ and $x \geq -3$ as a compound inequality without using *and*.**

$x < 0$ and $x \geq -3$ can be written as ____ $> x \geq$ ____ or ____ $\leq x <$ ____.

Your Turn Write $x > -2$ and $x \leq 5$ as a compound inequality without using *and*.

EXAMPLE

2 **A veterinarian has a scale for weighing dogs and cats that weigh more than 10 pounds *but no more than 65 pounds*. The weights w that can be measured on this scale can be written as $10 < w \leq 65$. Graph the solution of this inequality.**

Rewrite the compound inequality using ____. $10 < w \leq 65$ is the same as $w > 10$ and ____.

STEP 1 Graph $w >$ ____.

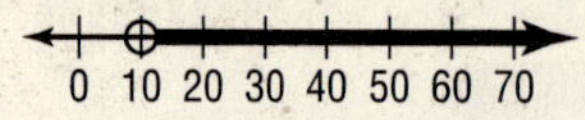

STEP 2 Graph $w \leq$ ____.

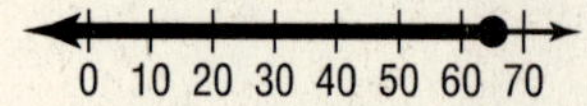

STEP 3 Find their ____.

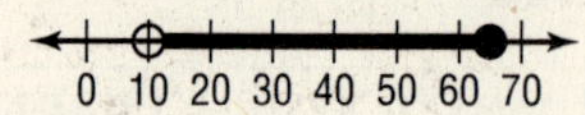

The solution is $\{w \mid$ ____ $< w \leq$ ____ $\}$.

Your Turn An amusement park charges \$15 admission for children between the ages of 5 and 13. The age a of people charged \$15 for admission can be written as $5 \leq a \leq 13$. Graph the solution of this inequality.

____

EXAMPLE

3 Solve $-6 \leq x - 3 \leq -1$. Graph the solution.

STEP 1 Rewrite the compound inequality using *and*.

$$-6 \leq x - 3 \leq -1$$

____ ≥ -6 and ____ ≤ -1

STEP 2 Solve each inequality.

$x - 3 \geq -6$ and $x - 3 \leq -1$

$x - 3 +$ ____ $\geq -6 +$ ____ $\qquad x - 3 +$ ____ $\leq -1 +$ ____

$x \geq$ ____ $\qquad x \leq$ ____

STEP 3 Rewrite the inequality as ____ $\leq x \leq$ ____.

The solution is $\{x \mid$ ____ $\}$.

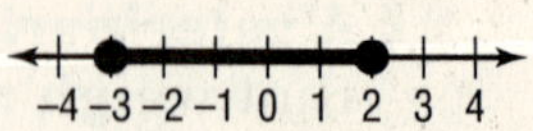

Your Turn Solve $4 < -2x < 12$. Graph the solution.

____

EXAMPLE

4 Graph the solution of $x \geq 7$ or $x < 3$.

STEP 1 Graph $x \geq$ ____.

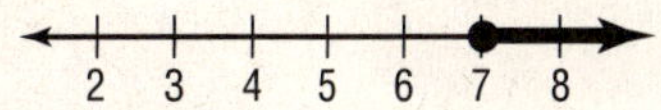

STEP 2 Graph $x <$ ____.

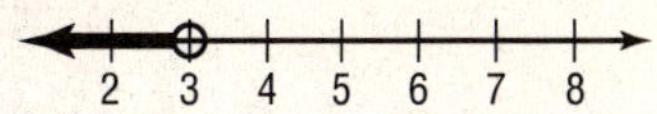

STEP 3 Find the union of the graphs.

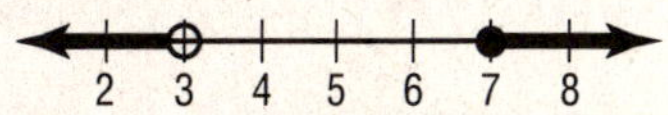

Your Turn Graph the solution of $x < -3$ or $x \geq 2$.

EXAMPLE

5 Solve $\frac{2}{3}x > 4$ or $-5x < 20$. Graph the solution.

$\frac{2}{3}x > 4$ or $-5x < 20$

____ $\left(\frac{2}{3}x\right) >$ ____ (4) $\qquad \frac{-5x}{\text{____}} > \frac{20}{\text{____}}$

$x >$ ____ $\qquad x >$ ____

Now graph the solution.

STEP 1 Graph $x >$ ____.

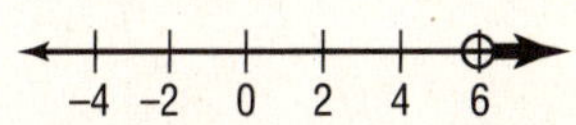

STEP 2 Graph $x > -4$.

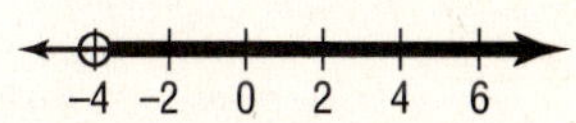

STEP 3 Find the union of the graphs.

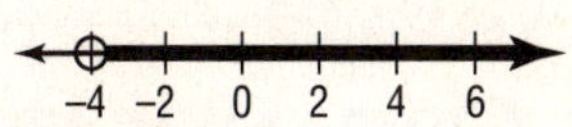

The last graph shows the solution $\{x \mid$ ____ $\}$.

HOMEWORK ASSIGNMENT

Page(s):

Exercises:

Your Turn Solve $\frac{1}{4}x > 3$ or $-3x < -48$. Graph the solution.

12–6 Solving Inequalities Involving Absolute Value

What You'll Learn

- Solve inequalities involving absolute value.

EXAMPLE

1 Solve $|x + 3| < 4$. Graph the solution.

CASE 1

$x + 3$ is positive.

$x + 3 <$ ____

$x + 3 -$ ____ $< 4 -$ ____

$x <$ ____

CASE 2

$x + 3$ is negative.

$-(x + 3) <$ ____

$x + 3 >$ ____

$x + 3 -$ ____ $> -4 -$ ____

$x >$ ____

The solution is $\{x \mid$ ____ $< x <$ ____ $\}$.

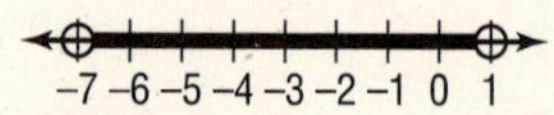

The solution makes sense since ____ and ____ are at most ____ units from ____.

Your Turn Solve $|t - 3| < 10$. Graph the solution.

EXAMPLE

Foldables™

Organize It

Summarize the difference between the solution of $|x| < 4$ and $|x| > 4$ under the tab for Lesson 12–6.

2 Solve $|4x| \geq 16$. Graph the solution.

CASE 1

$4x$ is positive.

$4x \geq$ ____

$\frac{4x}{4} \geq \frac{16}{4}$

$x \geq$ ____

CASE 2

$4x$ is negative.

$-4x \geq$ ____

$4x \leq$ ____

$\frac{4x}{4} \leq \frac{-16}{4}$

$x \leq$ ____

The solution is $\{x \mid x \leq$ ____ or $x \geq$ ____ $\}$.

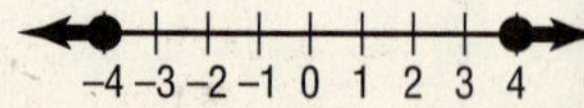

REMEMBER IT

The word "and" refers to an intersection of solutions and the word "or" refers to a union of solutions.

Your Turn Solve $|5x| \geq 15$. Graph the solution.

EXAMPLE

3 A lumber company makes $3\frac{1}{4}$-foot railing posts to use in making decks. The tolerance for the posts is 0.025 feet. What is the range of acceptable post lengths?

Let p = the actual measure of the posts.

Then, $\left|p - 3\frac{1}{4}\right| \leq$ ______.

$|p - 3.25| \leq 0.025$ Write $\frac{1}{4}$ as 0.25.

CASE 1

$p - 3.25$ is positive.

$p - 3.25 \leq$ ______

$p - 3.25 +$ ______ $\leq 0.025 +$ ______

$p \leq$ ______

CASE 2

$p - 3.25$ is negative.

$-(p - 3.25) \leq$ ______

$p - 3.25 \geq$ ______

$p - 3.25 +$ ______ $\geq -0.025 +$ ______

$p \geq$ ______

The solution is $\{p \mid$ ______ $\leq p \leq$ ______ $\}$.

HOMEWORK ASSIGNMENT

Page(s):

Exercises:

Your Turn A company makes $1\frac{1}{2}$-inch washers. The tolerance for the washers is 0.05 inch. What is the range of acceptable washer sizes?

12–7 Graphing Inequalities in Two Variables

WHAT YOU'LL LEARN

- Graph inequalities on the coordinate plane.

BUILD YOUR VOCABULARY (page 246)

A line that ______ the coordinate plane into half-planes is a **boundary**.

The region of the graph of an inequality on ______ side of a boundary is a **half-plane**.

WRITE IT

Describe the term "half-plane" using the words "half" and "plane".

______

______

______

______

______

______

EXAMPLE

1 Graph $y \leq x - 2$.

STEP 1 Determine the boundary by graphing the related equation, $y = x - 2$.

x	$x - 2$	y
-2	$-2 - 2$	
-1	$-1 - 2$	
0	$0 - 2$	
1	$1 - 2$	
2	$2 - 2$	

STEP 2 Draw a ______ line since the boundary is included.

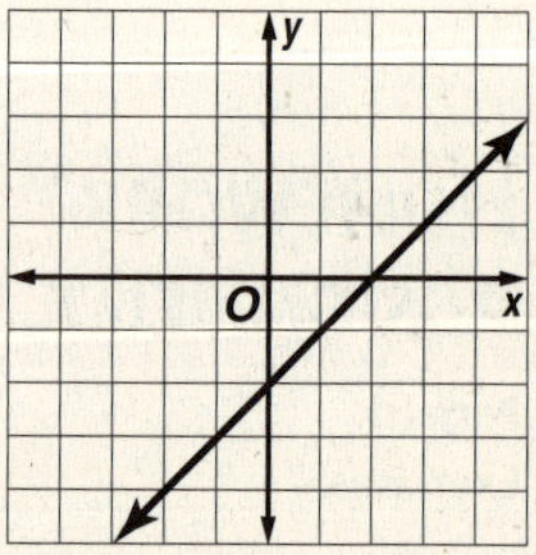

STEP 3 Test any point to find which half-plane is the solution. Use (0, 0) since it is the easiest point to use in calculations.

$y \leq x - 2$

$\square \overset{?}{\leq} \square - 2$ $x = \square$, $y = \square$

$0 \leq \square$ false

Since (0, 0) does not result in a ______ inequality, the half-plane containing (0, 0) is not the solution.

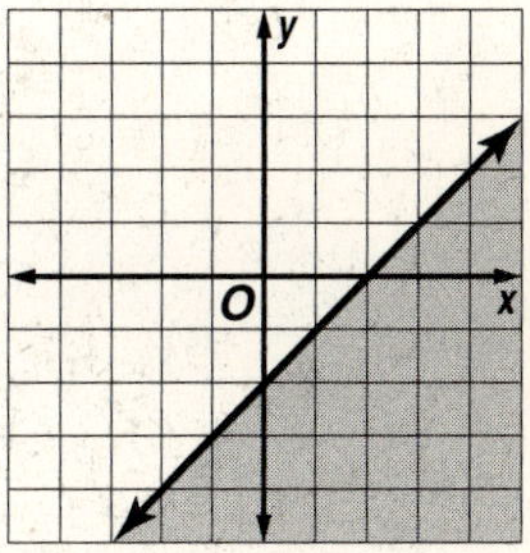

FOLDABLES™

ORGANIZE IT

Explain when to use a dashed line and when to use a solid line when graphing inequalities in two variables under the tab for Lesson 12–7. Then explain how to determine which half-plane should be shaded.

Your Turn Graph $y \geq -x - 2$.

EXAMPLE

2 **Graph $-3x + y > 1$.**

To make a table or graph for the ______ line, solve the inequality for y in terms of ______.

$-3x + y > 1$

$-3x + y + \square > 1 + \square$ Add ______ to each side.

$y > \square + 1$ Rewrite ______ as $3x + 1$.

STEP 1 Determine the boundry by graphing the related equation, $y = 3x + 1$

x	$3x + 1$	y
-2	$3(-2) + 1$	
-1	$3(-1) + 1$	
0	$3(0) + 1$	
1	$3(1) + 1$	
2	$3(2) + 1$	

STEP 2 Draw a ______ line since the boundary is not included.

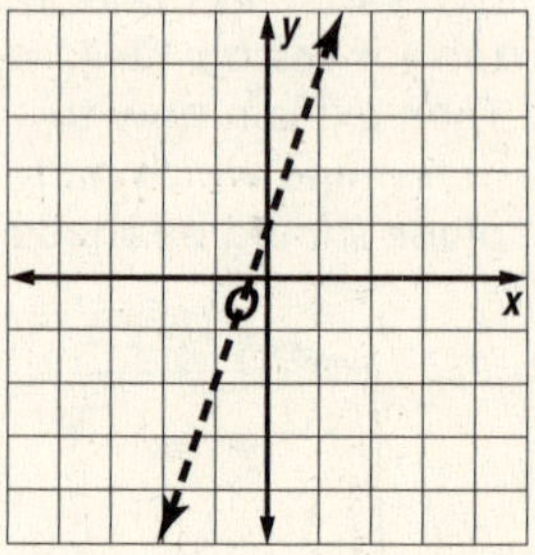

STEP 3 Test (0, 0) to find which half-plane contains the solution.

$y > 3x + 1$

$\square > 3(\square) + 1$ $x =$ ______, $y =$ ______

$0 > 1$ false

Since (0, 0) does not result in a true inequality, the half-plane containing (0, 0) is not the solution. Thus, shade the other half-plane.

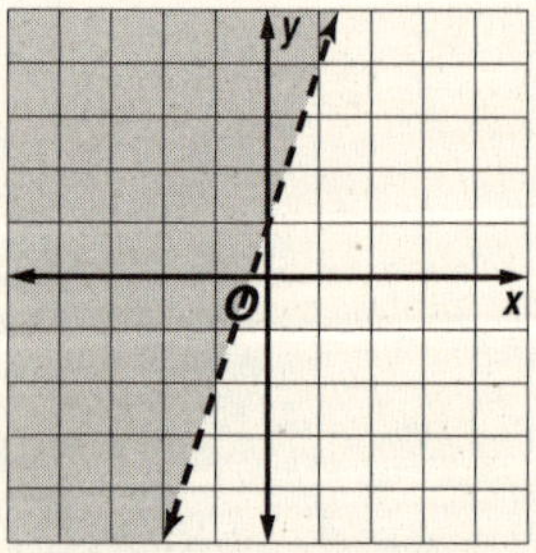

Your Turn Graph $y - 2x \leq 1$.

HOMEWORK ASSIGNMENT

Page(s):

Exercises:

CHAPTER 12

BRINGING IT ALL TOGETHER

STUDY GUIDE

FOLDABLES	VOCABULARY PUZZLEMAKER	BUILD YOUR VOCABULARY
Use your **Chapter 12 Foldable** to help you study for your chapter test.	To make a crossword puzzle, word search, or jumble puzzle of the vocabulary words in Chapter 12, go to: www.glencoe.com/sec/math/t_resources/free/index.php	You can use your completed **Vocabulary Builder** (page 246) to help you solve the puzzle.

12-1 Inequalities and Their Graphs

Write the letter of the graph that matches each inequality.

1. $x \leq -1$

2. $x \geq -1$

3. $x < -1$

4. $x > -1$

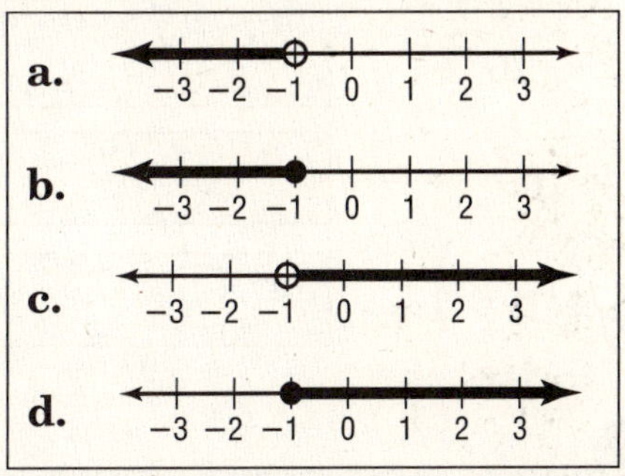

12-2 Solving Addition and Subtraction Inequalities

Write an inequality for each statement. Then solve.

5. A number subtracted from 21 is no less than −2.

6. A number added to 12 is a minimum of −1.

7. 5 more than a number is at least 15.

8. 18 less than a number is at most 45.

12-3 Solving Multiplication and Division Inequalities

Solve each inequality. Check your solution.

9. $12 < 6n$

10. $\frac{t}{-3} \geq 14$

11. $12x \leq 32$

12-4 Solving Multi-Step Inequalities

Solve each inequality. Check your solution.

12. $2x - 8 \geq -16$

13. $3y - 5 < 16$

14. $n - 3(n + 1) > 1$

12-5 Solving Compound Inequalities

Write the letter of the graph that matches each compound inequality.

15. $x > 3$ or $x < -3$

16. $-3 < x < 3$

17. $x > -3$ or $x > 3$

a.
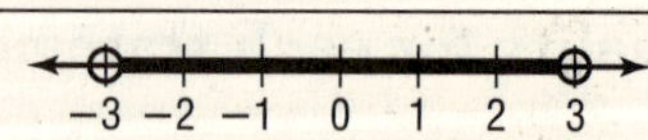

b.
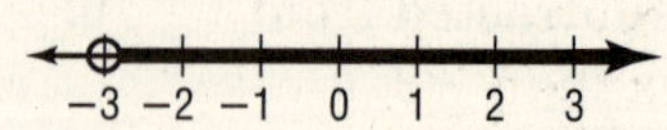

c.
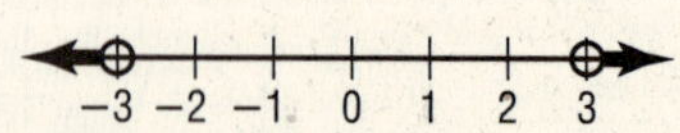

Solve each compound inequality. Graph the solution.

18. $-13 \leq 2x - 1 \leq -5$

19. $3a \geq 21$ or $-2a \leq -24$

12-6

Solving Inequalities Involving Absolute Value

Solve each inequality. Graph the solution.

20. $|2x + 2| \geq 8$

$2x + 2 \geq 8$

$2x + 2 \leq -8$

21. $|x - 5| \leq 4$

$x - 5 \leq 4$

$x - 5 \geq -4$

22. $|2x - 3| > 5$

$2x - 3 > 5$

$2x - 3 < -5$

23. $|2x - 2| \leq 6$

$2x - 2 \leq 6$

$2x - 2 \geq -6$

12-7

Graphing Inequalities in Two Variables

24. **BUSINESS** A small business is charged 5 cents per minute for in-state calls and 10 cents per minute for out of state calls. The inequality $5x + 10y < 2500$ represents how many minutes of each type of call can be made for under \$25 per month. Graph the inequality. List three solutions of the inequality.

CHAPTER 12

Checklist

ARE YOU READY FOR THE CHAPTER TEST?

Math Online

Visit **algconcepts.com** to access your textbook, more examples, self-check quizzes, and practice tests to help you study the concepts in Chapter 12.

Check the one that applies. Suggestions to help you study are given with each item.

- [ ] **I completed the review of all or most lessons without using my notes or asking for help.**
  - You are probably ready for the Chapter Test.
  - You may want take the Chapter 12 Practice Test on page 545 of your textbook as a final check.

- [ ] **I used my Foldable or Study Notebook to complete the review of all or most lessons.**
  - You should complete the Chapter 12 Study Guide and Review on pages 542–544 of your textbook.
  - If you are unsure of any concepts or skills, refer back to the specific lesson(s).
  - You may also want to take the Chapter 12 Practice Test on page 545.

- [ ] **I asked for help from someone else to complete the review of all or most lessons.**
  - You should review the examples and concepts in your Study Notebook and Chapter 12 Foldable.
  - Then complete the Chapter 12 Study Guide and Review on pages 542–544 of your textbook.
  - If you are unsure of any concepts or skills, refer back to the specific lesson(s).
  - You may also want to take the Chapter 12 Practice Test on page 545.

Student Signature

Parent/Guardian Signature

Teacher Signature

Systems of Equations and Inequalities

FOLDABLES™ Use the instructions below to make a Foldable to help you organize your notes as you study the chapter. You will see Foldable reminders in the margin of this Interactive Study Notebook to help you in taking notes.

Begin with four sheets of grid paper.

STEP 1 **Stack**
Stack sheets of paper with edges four grids apart to create tabs.

STEP 2 **Fold**
Fold up bottom edges. All tabs should be the same size.

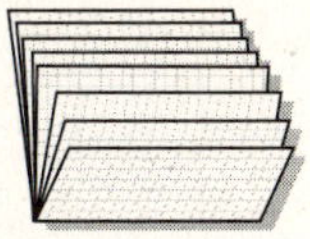

STEP 3 **Staple**
Staple along the fold.

STEP 4 **Label**
Label the tabs using lesson numbers and titles.

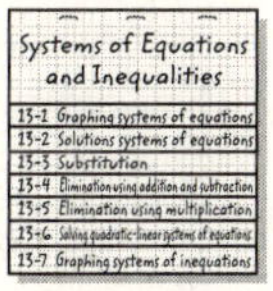

NOTE-TAKING TIP: When taking notes, think about the order in which concepts are being presented. Write why you think the concepts were presented in this sequence.

CHAPTER 13

BUILD YOUR VOCABULARY

This is an alphabetical list of new vocabulary terms you will learn in Chapter 13. As you complete the study notes for the chapter, you will see Build Your Vocabulary reminders to complete each term's definition or description on these pages. Remember to add the textbook page number in the second column for reference when you study.

Vocabulary Term	Found on Page	Definition	Description or Example
augmented matrix			
consistent [kun-SIS-tunt]			
dependent			
digit problems			
elimination [ee-LIM-in-AY-shun]			
identity matrix			

Vocabulary Term	Found on Page	Definition	Description or Example
inconsistent [in-kun-SIS-tunt]			
independent			
matrices [MAY-tra-seez]			
quadratic-linear system of equations			
row operations			
substitution [SUB-sti-TOO-shun]			
system of equations			
system of inequalities			

13–1 Graphing Systems of Equations

WHAT YOU'LL LEARN

- Solve systems of equations by graphing.

KEY CONCEPT

System of Equations
A system of equations is a set of two or more equations with the same variables. The solution is the ordered pair that satisfies all of the equations.

EXAMPLES

Solve each system of equations by graphing.

1 $y = x + 6$
$y = 3x$

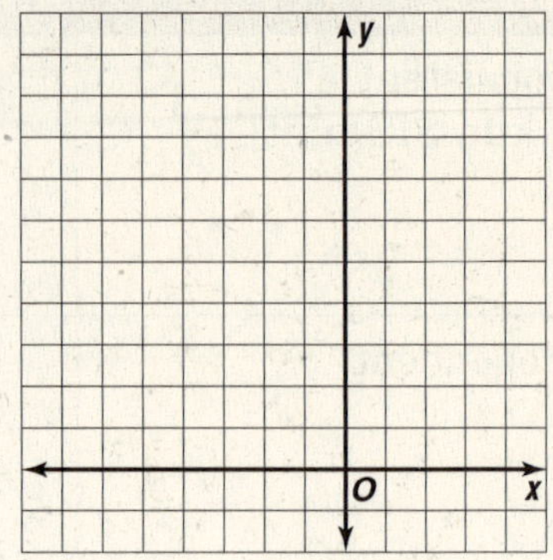

The graphs appear to intersect at ______.

You can check this estimate by substituting the coordinates into each equation. The solution of the system of equations is ______.

2 $x - y = 3$
$x + y = 1$

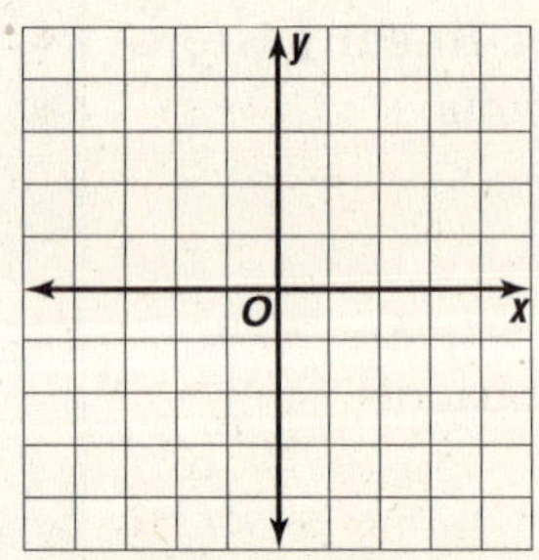

The graphs appear to intersect at ______.

Check this estimate.

The solution of the system of equations is ______.

Your Turn **Solve each system of equations by graphing.**

a. $y = x$
$x + y = 6$

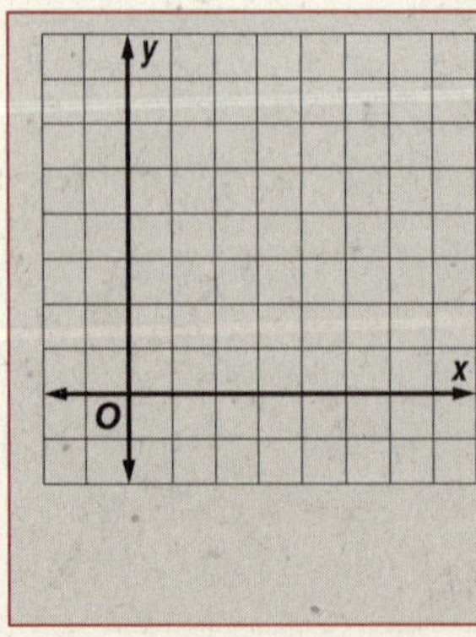

b. $y + x = 4$
$y - x = 6$

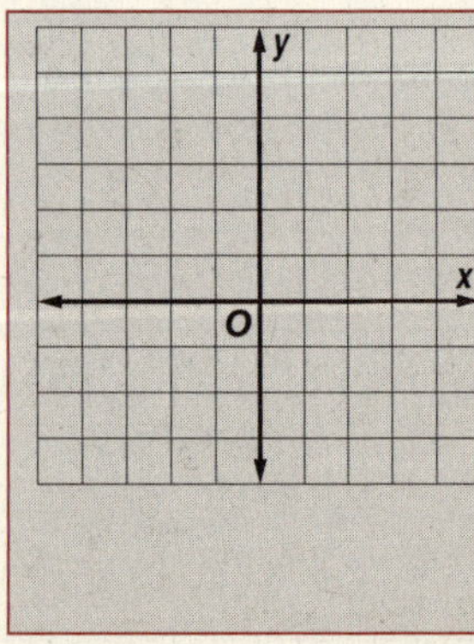

EXAMPLE

3 The Math Club is selling T-shirts for a profit of $4 each and caps for a profit of $5 each. The club wants to sell 50 items and make a profit of $240. How many of each item should the club try to sell?

Let x = the number of T-shirts and y = the number of caps. You can write two equations to represent this situation.

$x + y = \square$ the number of items

$4x + 5y = \square$ the total profit

Graph $x + y = \square$ and $4x + 5y = \square$. The graphs appear to intersect at ($\square$, 40). Check this estimate.

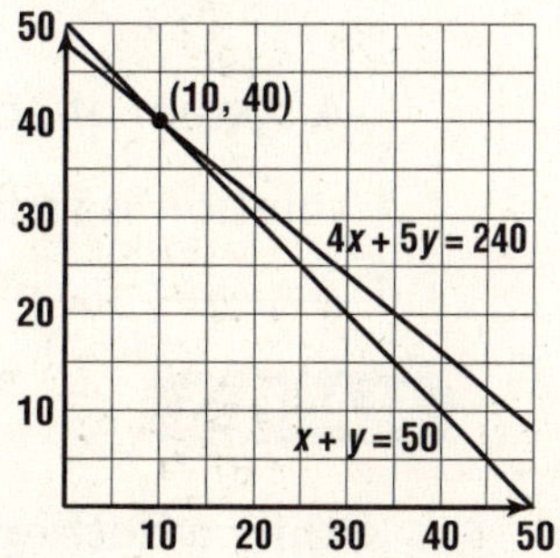

$x + y = 50$

$\square + \square = 50$

$\square = 50$ ✓

$4x + 5y = 240$

$4(\square) + 5(\square) = 240$

$\square + \square = 240$

$\square = 240$ ✓

They should try to sell $\square$ T-shirts and $\square$ caps.

Your Turn A service organization is selling flowers for a profit of $2 each and vegetable plants for a profit of $3 each. The club wants to sell 100 items and make a profit of $230. How many of each item should the organization try to sell?

FOLDABLES™

ORGANIZE IT

Tell what a system of equations is under the tab for Lesson 13-1. Then describe the solution to a system of equations.

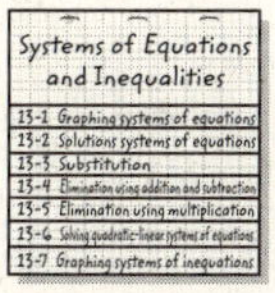

HOMEWORK ASSIGNMENT

Page(s):

Exercises:

13–2 Solutions of Systems of Equations

What You'll Learn

- Determine whether a system of equations has one solution, no solution, or infinitely many solutions by graphing.

Build Your Vocabulary (pages 270–271)

A system of equations with ______ solution is **consistent**.

A system of equations with ______ solutions is **dependent**.

A system of equations with ______ is **inconsistent**.

A system of equations with ______ solution is **independent**.

Examples

State whether each system is consistent and independent, consistent and dependent, or inconsistent.

1

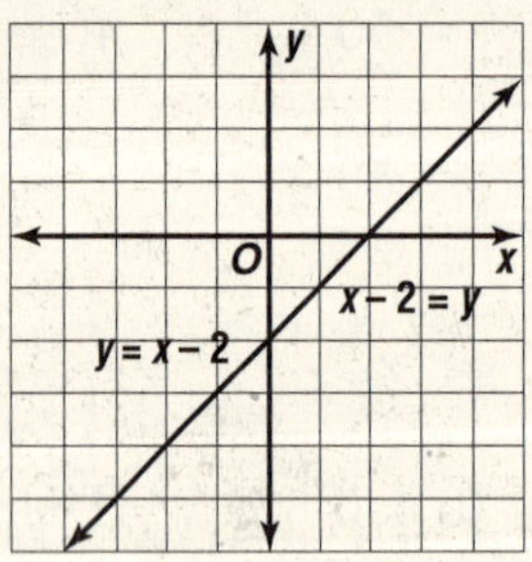

Both equations have the same graph. Because any ordered pair on the graph will satisfy both equations, there are ______ solutions. The system is ______ and ______.

2

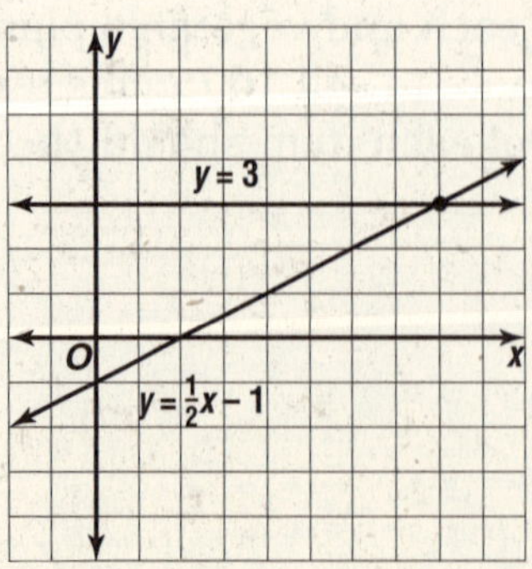

The graphs appear to intersect at the point at ______. Because there is ______ solution, this system of equations is ______ and ______.

Write It

How do you know if two lines are parallel without drawing their graphs? How do you know, without graphing, when two equations represent the same line?

______

______

______

______

______

Your Turn **State whether each system is consistent and independent, consistent and dependent, or inconsistent.**

a.

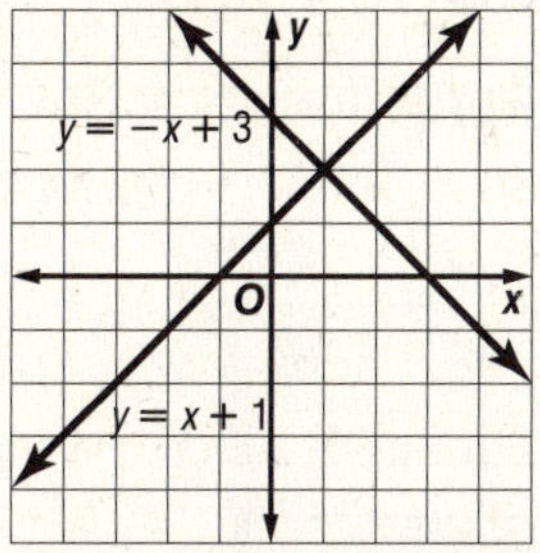

b.

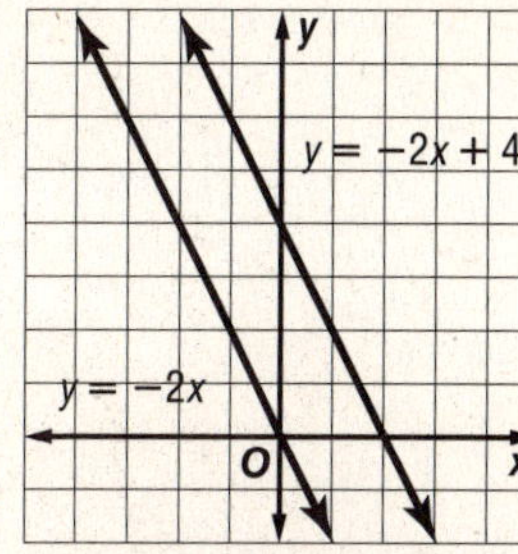

EXAMPLES

Determine whether the system of equations has one solution, no solution, or infinitely many solutions by graphing. If the system has one solution, name it.

3 $3x - y = 1$

$6x - 2y = 2$

Write each equation in slope-intercept form.

$3x - y = 1 \longrightarrow y =$ ______

$6x - 2y = 2 \longrightarrow y =$ ______

The graphs have the same ______ and the same ______. The system of equations has ______ solutions.

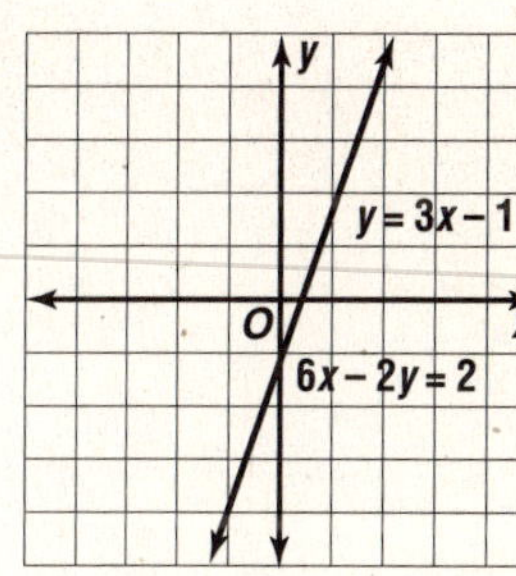

4 $y = -x$

$y = x - 2$

The graphs appear to intersect at ______. Therefore, this system of equations has one solution ______.

Remember to check by substituting the values for x and y into each of the original equations.

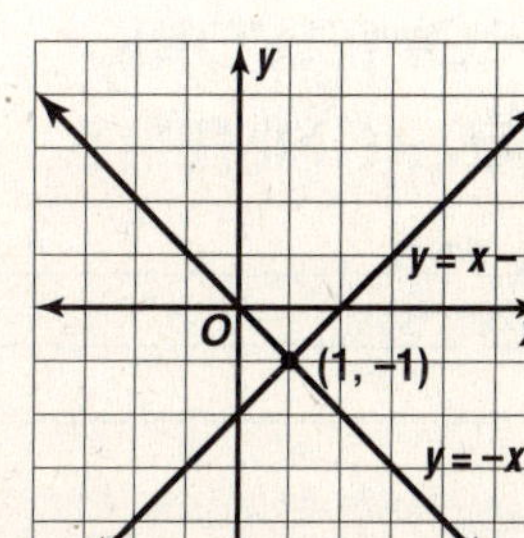

FOLDABLES

ORGANIZE IT

Describe the possible graphs of a system of linear equations under the tab for Lesson 13-2. Then name the type of system that each graph represents.

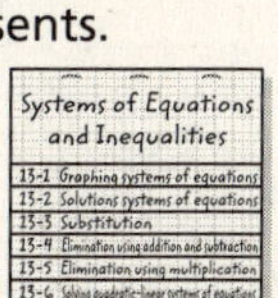

Your Turn **Determine whether the system of equations has one solution, no solution, or infinitely many solutions by graphing. If the system has one solution, name it.**

a. $y = 3x - 1$
$y = 2$

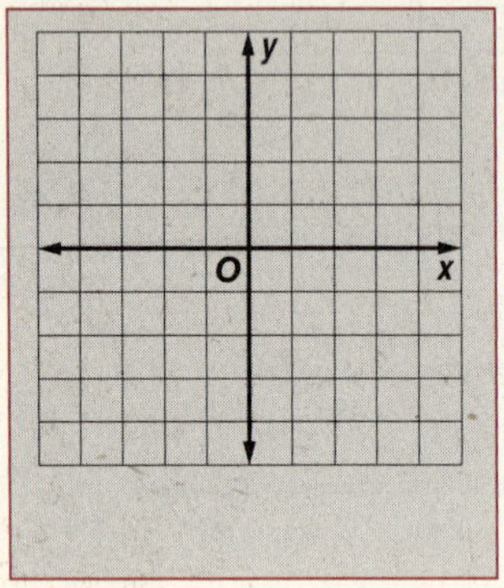

b. $2x - 3y = 1$
$4x - 6y = 5$

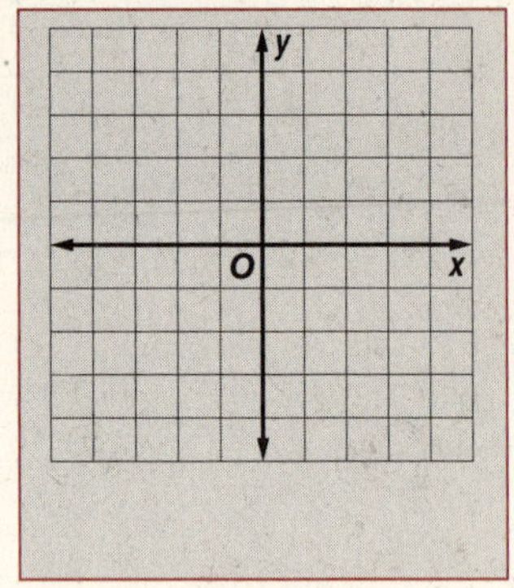

EXAMPLE

5 The system of equations $4x + 2y = 6$ and $8x + 4y = 20$ represents the tracks of two trains. Do the tracks intersect, run parallel, or are the trains running on the same track? Explain.

Write each equation in slope-intercept form.

$4x + 2y = 6 \longrightarrow y =$ ______

$8x + 4y = 20 \longrightarrow y =$ ______

The graphs have the same ______ and ______ y-intercepts, so the lines are ______ and the tracks do not intersect.

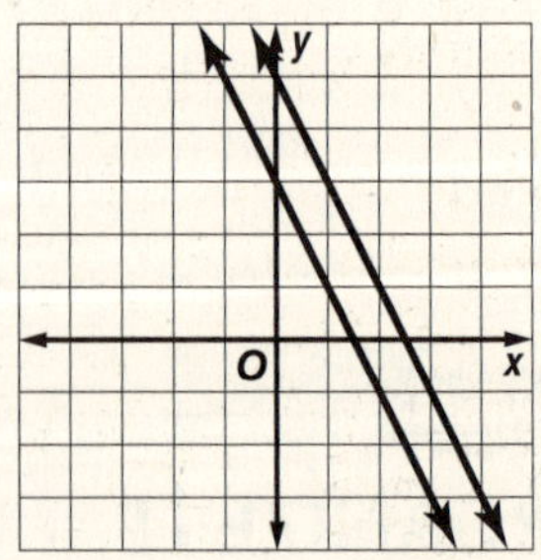

Your Turn The system of equations $3x + y = 2$ and $9x + 3y = 6$ represents the tracks of two trains. Do the tracks intersect, run parallel, or are the trains on the same track? Explain.

HOMEWORK ASSIGNMENT

Page(s):

Exercises:

13-3 Substitution

What You'll Learn

- Solve systems of equations by the substitution method.

Build Your Vocabulary (page 271)

Substitution is an algebraic method to solve a of equations.

Foldables

Organize It

Summarize the substitution method and explain when it is preferred over the graphing method under the tab for Lesson 13-3.

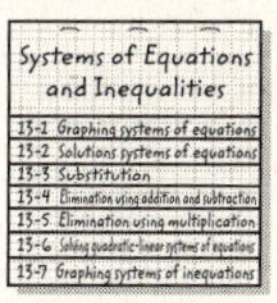

Example

1 Use substitution to solve $x = 2y$ and $2x - 3y = 5$.

The first equation tells you that x is equal to $2y$. So, substitute $2y$ for x in the second equation. Then solve for y.

$2x - 3y = 5$

$2(\quad) - 3y = 5$ Replace x with $\quad$.

$\quad - 3y = 5$

$\quad = 5$

Now substitute 5 for y in either equation and solve for x. Choose the equation that is easier to solve.

$x = 2y$

$x = 2(\quad)$ or $\quad$ Replace y with $\quad$.

The solution of this system of equations is $\quad$.

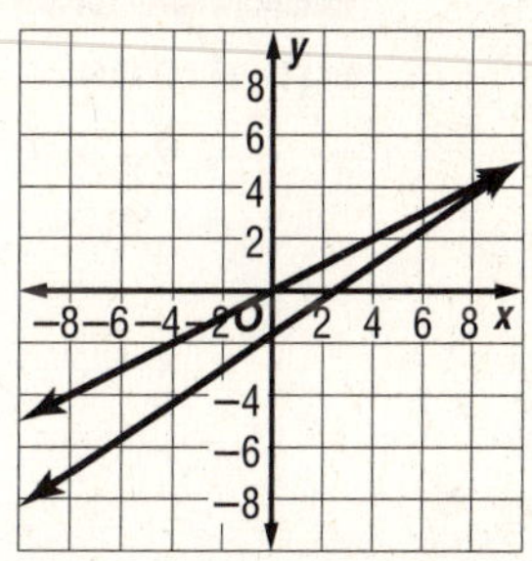

Your Turn Use substitution to solve $y = 3x + 1$ and $x - y = -3$.

EXAMPLE

2 **Use substitution to solve $-2x + y = -7$ and $3x - 2y = 12$.**

Solve the first equation for y since the coefficient of y is ______.

$-2x + y = -7 \longrightarrow y = 2x - 7$

Next, find the value of x by substituting $2x - 7$ for y in the second equation.

$3x - 2y = 12$

$3x - 2(\text{______}) = 12$

$3x - \text{______} + \text{______} = 12$

$\text{______} + 14 = 12$

$-x + 14 - \text{______} = 12 - \text{______}$

$-x = \text{______}$

$\dfrac{-x}{\text{______}} = \dfrac{-}{\text{______}}$

$x = \text{______}$

Now substitute ______ for x in either equation and solve for y.

$-2(\text{______}) + y = -7$

$\text{______} + y = -7$

$-4 + y + \text{______} = -7 + \text{______}$

$y = \text{______}$

The solution is ______.

Your Turn **Use substitution to solve each system of equations.**

a. $x - 2y = 0$
$3x - 2y = 12$

b. $2x - y = 3$
$x - 2y = 6$

EXAMPLES

Use substitution to solve each system of equations.

REMEMBER IT When using an algebraic method to solve a system of equations, a true statement with no variables represents an infinite number of solutions. A false statement with no variables represents no solution.

3 $\mathbf{3x + 1 = y}$
$\mathbf{9x - 3y = -3}$

Find the value of x by substituting ________ for y in the second equation.

$9 - 3y = -3$

$9x - 3(\_\_\_\_) = -3$ Replace y with ________.

$9x - \_\_\_\_ - \_\_\_\_ = -3$ Distributive Property

$\_\_\_\_ = -3$

The statement $-3 = -3$ is ________. This means that an ordered pair for any point on either line is a solution to both equations. The system has ________ solutions.

4 $\mathbf{y = 3x - 5}$
$\mathbf{6x - 2y = 4}$

$6x - 2(\_\_\_\_) = 4$ Replace y with ________.

$6x - \_\_\_\_ + \_\_\_\_ = 4$ Distributive Property

$\_\_\_\_ = 4$

The statement $10 = 4$ is ________. This means that there are no ordered pairs that are solutions to both equations. Thus, the lines are ________, and the system has ________.

Your Turn **Use substitution to solve each system of equations.**

a. $x - 2y = 2$
$3x - 6y = 12$

b. $y = 2x - 2$
$2x - y = 2$

HOMEWORK ASSIGNMENT

Page(s):

Exercises:

13–4 Elimination Using Addition and Subtraction

WHAT YOU'LL LEARN

- Solve systems of equations by the elimination method using addition and subtraction.

BUILD YOUR VOCABULARY (page 270)

Elimination is an algebraic method to solve a system of equations by adding or subtracting the equations.

REVIEW IT

How can you tell when a linear equation is in standard form? *(Lesson 6-3)*

____

____

____

____

____

____

____

____

EXAMPLE

1 Use elimination to solve $3x - 2y = 21$ and $3x + 4y = 3$.

$$3x - 2y = 21$$
$$(-)3x + 4y = 3$$

Write the equations in column form. Subtract to eliminate the x terms.

$$\_\_\_\_ - \_\_\_\_ = 18$$

$$\_\_\_\_ = 18$$

$$\frac{-6y}{-6} = \frac{18}{-6}$$

$$y = \_\_\_\_$$

The value of y is ____.

Now substitute in either equation to find the value of x. Choose the equation that is easier for you to solve.

$$3x + 4(\_\_\_\_) = 3$$

$$3x - \_\_\_\_ = 3$$

$$3x - 12 + \_\_\_\_ = 3 + \_\_\_\_$$

$$3x = 15$$

$$\frac{3x}{3} = \frac{15}{3}$$

$$x = \_\_\_\_$$

The value of x is ____.

The solution of the system of equations is ____.

Your Turn Use elimination to solve $x + 2y = 4$ and $x - y = 1$.

EXAMPLE

2 **For a special event at the Kartchner Caverns, the cost of tickets for 4 adults and 3 students was \$84 and the cost of tickets for 2 adults and 3 students was \$54. Find the admission price for an adult and for a student at the special event.**

Let a = the cost for an ______ and s = the cost for a ______.

$4a + 3s = 84$ total cost for the ______ group

$2a + 3s = 54$ total cost for the ______ group

$4a + 3s = 84$ Write the equations in column form.

$(-)\quad 2a + 3s = 54$ Subtract the equations to eliminate the s terms.

$\_\_\_ + \_\_\_ = 30$

$\_\_\_ = 30$

$\frac{2a}{2} = \frac{30}{2}$

$a = \_\_\_$ The value of a is ______.

Now substitute in either equation to find the value of s.

$4a + 3s = 84$

$4(\_\_\_) + 3s = 84$

$\_\_\_ + 3s = 84$

$60 + 3s - \_\_\_ = 84 - \_\_\_$

$3s = 24$

$\frac{3s}{3} = \frac{24}{3}$

$s = \_\_\_$ The value of s is ______.

The solution of the system of equations is ______. This means that the cost for adults was ______ and the cost for students was ______.

Your Turn The group admission cost for 2 adults and 6 children to enter a museum is \$60. The group admission cost for 2 adults and 10 children to enter the same museum is \$84. Find the group admission price for an adult and a child.

___

EXAMPLE

3 Use elimination to solve $3x + 4y = 6$ and $5x - 4y = -22$.

$$\begin{aligned} 3x + 4y &= 6 \\ (+)5x - 4y &= -22 \end{aligned}$$

$4y$ and $-4y$ are additive inverses. Add to eliminate the y terms.

$$\underline{\quad} + \underline{\quad} = -16$$

$$\underline{\quad} = -16$$

$$\frac{8x}{8} = \frac{-16}{8x}$$

$x = \underline{\quad}$ The value of x is ___.

Now substitute in either equation to find the value of y.

$$3x + 4y = 6$$

$$3(\underline{\quad}) + 4y = 6$$

$$\underline{\quad} + 4y = 6$$

$$-6 + 4y + \underline{\quad} = 6 + \underline{\quad}$$

$$4y = 12$$

$$\frac{4y}{4} = \frac{12}{4}$$

$y = \underline{\quad}$ The value of y is ___.

The solution of the system is ___.

FOLDABLES™

ORGANIZE IT

Describe the types of systems where the elimination method is the preferred method used to solve under the tab for Lesson 13-4.

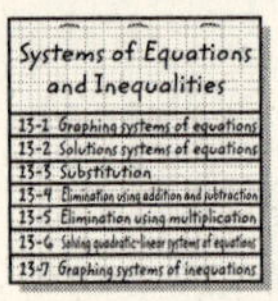

Your Turn Use elimination to solve $2x + 3y = -6$ and $3x - 3y = 21$.

___

BUILD YOUR VOCABULARY (page 270)

Digit problems explore the relationship between digits of a number.

EXAMPLE

4 The sum of the digits of a two-digit number is 14. If the units digit is 2 more than the tens digit, what is the number?

Let t represent the tens digit and let u represent the units digit.

$t + u = 14$ ⟶ the sum of the digits
$u = t + 2$ ⟶ the relationship between the digits

Rewrite the second equation so that the t and u are on the same side of the equation.

$u = t + 2$ ⟶ $-t + u = 2$

Then use elimination to solve.

$t + u = 14$ Write the equations in column form.
$(+) -t + u = 2$ Add the equations to eliminate the t terms.

☐ + ☐ = 16
☐ = 16

$\frac{2u}{2} = \frac{16}{2}$

$u =$ ☐ The units digit is ☐.

Now substitute to find the tens digit.

$t + u = 14$
$t +$ ☐ $= 14$
$t + 8 - 8 = 14 - 8$
$t =$ ☐ The tens digit is ☐.

Since t is ☐ and u is ☐, the number is ☐.

HOMEWORK ASSIGNMENT
Page(s):
Exercises:

Your Turn The sum of the digits of a two-digit number is 7. If the units digit is 7 less than the tens digit, what is the number?

13–5 Elimination Using Multiplication

WHAT YOU'LL LEARN

- Solve systems of equations by the elimination method using multiplication and addition.

EXAMPLES

1 **Use elimination to solve $5x + 2y = -2$ and $3x - 6y = -30$.**

Multiply the first equation by 3 so that the y terms are additive inverses.

$$5x + 2y = -2 \xrightarrow{\text{Multiply by 3.}} \_\_\_\_x + \_\_\_\_y = -6$$

$$3x - 6y = -30 \qquad (+)\ 3x - 6y = -30$$

$$\_\_\_\_ + 0 = \_\_\_\_$$

$$\_\_\_\_ = -36$$

$$\frac{18x}{\_\_\_\_} = \frac{-36}{\_\_\_\_}$$

$$x = \_\_\_\_$$

Now find the value of y by replacing x in either equation.

$$5x + 2y = -2$$

$$5(\_\_\_\_) + 2y = -2$$

$$-10 + 2y = -2$$

$$-10 + 2y + \_\_\_\_ = -2 + \_\_\_\_$$

$$2y = \_\_\_\_$$

$$\frac{2y}{2} = \frac{8}{2}$$

$$y = \_\_\_\_$$

The solution of this system of equations is ____.

Your Turn Use elimination to solve $3x + 4y = -1$ and $2x + 8y = 0$.

FOLDABLES™

ORGANIZE IT

Summarize the difference between the elimination method taught in Lesson 13-4 and the elimination method taught in Lesson 13-5 under the tab for Lesson 13-5. Then give examples of when to use each method.

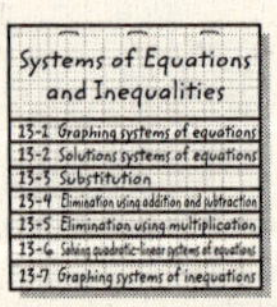

REMEMBER IT

Another method can be used to check your answers. For example, if elimination is used to solve the system, the system can then be graphed or substitution can be used to check the answer.

2 Use elimination to solve $3x + 2y = 4$ and $9x - 4y = 7$.

Multiply the first equation by -3 so that the x terms are additive inverses.

$3x + 2y = 4$ → Multiply by 3. → $\square x - \square y = \square$

$9x - 4y = 7$ → $(+)\ 9x - 4y = 7$

$$0 - \square = \square$$

$$\square = -5$$

$$\frac{-10y}{\square} = \frac{-5}{\square}$$

$$y = \square$$

Now find the value of x by replacing y in either equation.

$$3x + 2y = 4$$

$$3x + 2\left(\square\right) = 4$$

$$3x + \square = 4$$

$$3x + 1 - \square = 4 - \square$$

$$3x = \square$$

$$\frac{3x}{\square} = \frac{3}{\square}$$

$$x = \square$$

The solution of the system of equations is $\square$.

Your Turn Use elimination to solve $x - 2y = 3$ and $4x + y = 12$.

$\square$

EXAMPLE

3 **A music store has one price for all CDs except for the CDs in the Sale section. One customer bought 8 regular CDs and 2 sale CDs and paid $79.50. Another customer bought 4 regular CDs and 5 sale CDs and paid $62.75. What are the costs of regular and sale CDs?**

Let r represent the cost of the regular CDs and let s represent the cost of the sale CDs.

$8r + 2s = 79.50$ → first customer's expense

$4r + 5s = 62.75$ → second customer's expense

Multiply the second equation by -2 to eliminate the r terms.

$$8r + 2s = 79.50 \qquad\qquad 8r + 2s = 79.50$$

$$4r + 5s = 62.75 \xrightarrow[\text{by } -2.]{\text{Multiply}} (+)\ \square r - \square s = -125.50$$

$$0 - \square = \square$$

$$\square = -46$$

$$\frac{-8s}{\square} = \frac{-46}{\square}$$

$$s = \square$$

Now find the value of r by replacing s with ______ in either equation.

$$8r + 2s = 79.50$$

$$8r + 2(\square) = 79.50$$

$$8r + \square = 79.50$$

$$8r = \square$$

$$\frac{8r}{\square} = \frac{68}{\square}$$

$$r = \square$$

The solution is ______. This means that the regular CDs sell for ______ and the sale CDs sell for ______.

Your Turn Morgan and Amy found a special on shorts and shirts. Morgan bought 5 pairs of shorts and 5 shirts for $125, not including tax. Amy bought 3 pairs of shorts and 6 shirts for $105, not including tax. Find the cost of each pair of shorts and the cost of each shirt.

EXAMPLE

4 Use elimination to solve $4x - 3y = -8$ and $7x + 5y = 27$.

Multiply the first equation by 5 and the second equation by 3 so that the y terms are additive inverses.

$4x - 3y = -8$ Multiply by 5. → $\_\_x - \_\_y = \_\_$

$7x + 5y = 27$ Multiply by 3. → $(+)\ \_\_x + \_\_y = \_\_$

$41x = \_\_$

$\frac{41x}{41} = \frac{41}{41}$

$x = \_\_$

Now find the value of y by replacing x in either equation.

$7x + 5y = 27$

$7(\_\_) + 5y = 27$

$\_\_ + 5y = 27$

$7 + 5y - 7 = 27 - 7$

$5y = 20$

$\frac{5y}{5} = \frac{20}{5}$

$y = \_\_$ The solution is __.

REMEMBER IT

There may be more than one way to solve a problem. In Example 4, you can also solve this system of equations by multiplying the first equation by 7 and the second equation by −4.

HOMEWORK ASSIGNMENT

Page(s):

Exercises:

Your Turn Use elimination to solve $2x - 3y = 13$ and $3x + 4y = -6$.

13–6 Solving Quadratic-Linear Systems of Equations

EXAMPLES

WHAT YOU'LL LEARN

- Solve systems of quadratic and linear equations.

Determine whether each system of equations has one solution, two solutions, or no solution by graphing. If the system has one solution or two solutions, name them.

1 $y = x^2$
$y = -x$

The graphs appear to intersect at ______ and ______. Check this estimate by substituting the coordinates into each equation.

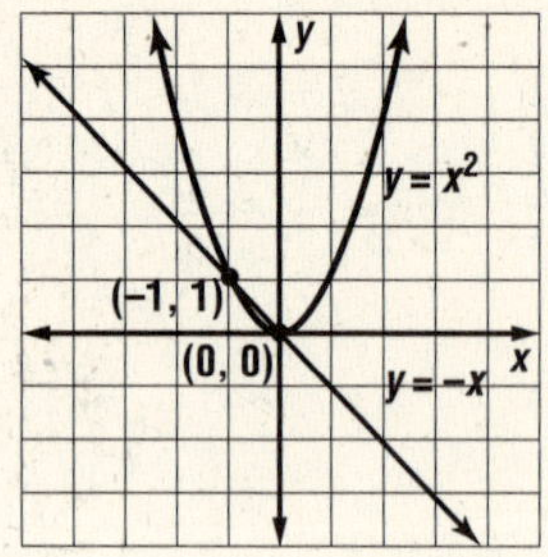

Check:

$y = x^2$	$y = -x$
$\_\_ \stackrel{?}{=} \_\_^2$ $(x, y) = \_\_\_\_$	$\_\_ \stackrel{?}{=} -\_\_$ $(x, y) = \_\_\_\_$
$\_\_ = 0$ ✓	$\_\_ \stackrel{?}{=} 0$ ✓

$y = x^2$	$y = -x$
$\_\_ \stackrel{?}{=} (\_\_)^2$ $(x, y) = \_\_\_\_$	$\_\_ = -(\_\_)$ $(x, y) = \_\_\_\_$
$\_\_ = 1$ ✓	$\_\_ = 1$ ✓

2 $y = -x^2 + 1$
$y = 1$

The graphs appear to intersect at ______.

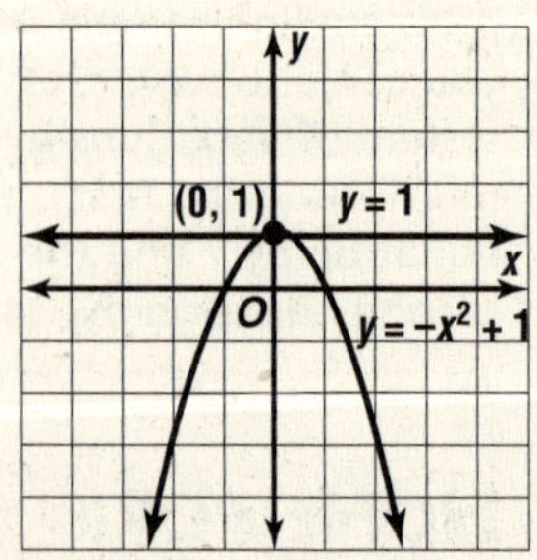

Check:

$y = -x^2 + 1$

$\_\_ \stackrel{?}{=} -(\_\_)^2 + 1$ $(x, y) = \_\_\_\_$

$1 = \_\_$ ✓

$y = 1$

$\_\_ = 1$ ✓

KEY CONCEPT

Quadratic-Linear Systems

no solution graphs do not intersect

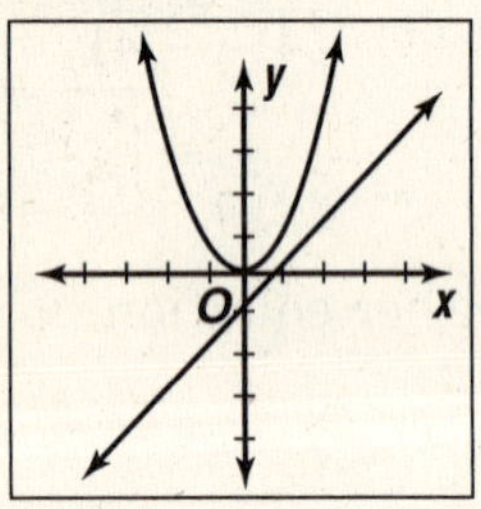

one solution graphs intersect at one point

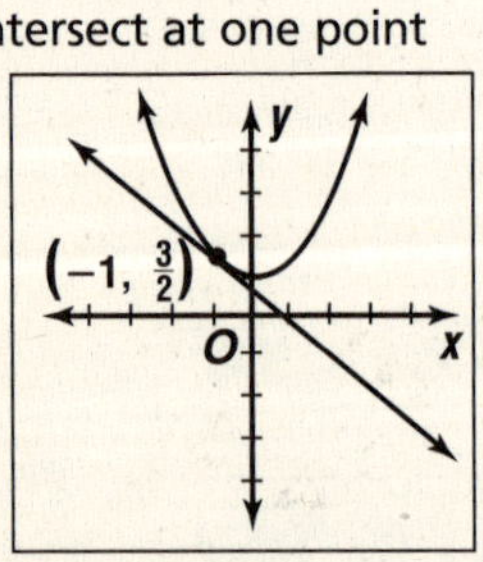

two solutions graphs intersect at two points

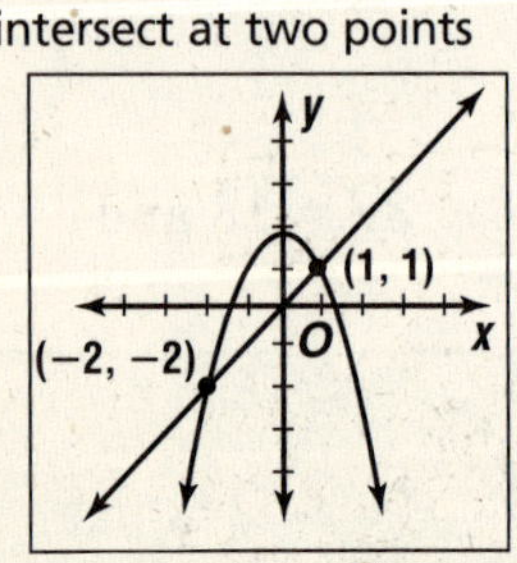

Your Turn **Determine whether each system of equations has one solution, two solutions, or no solution by graphing. If the system has one solution or two solutions, name them.**

a. $y = x^2 + 1$
$y = 5$

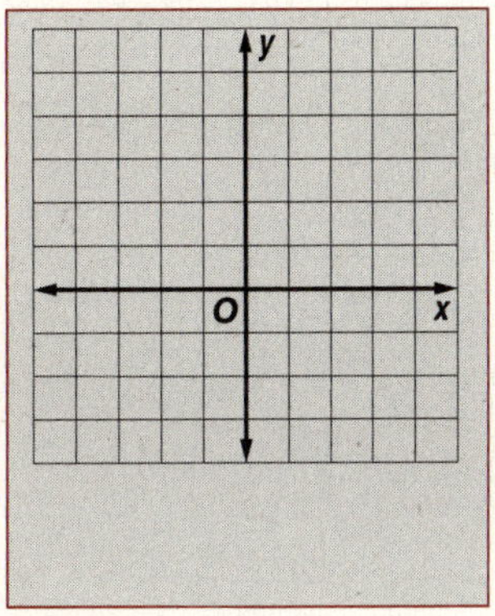

b. $y = -x^2 + 3$
$2x + y = 6$

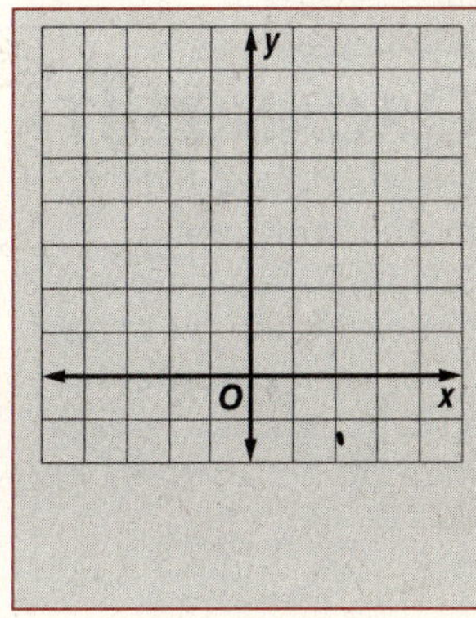

EXAMPLES

3 **Use substitution to solve $y = 2x^2 - 3$ and $y = -1$.**

Substitute -1 for y in the first equation. Then solve for x.

$y = 2x^2 - 3$

$\square = 2x^2 - 3$ Replace y with $\square$.

$-1 + \square = 2x^2 - 3 + \square$ Add $\square$ to each side.

$\square = 2x^2$

$\dfrac{2}{\square} = \dfrac{2x^2}{\square}$

$1 = x^2$

$\square = x$ Take the square root of each side.

The solutions of the system of equations are $\square$ and $\square$.

4 Use substitution to solve $y = 2x^2 + 3$ and $y = 1$.

Substitute 1 for y in the first equation.

$$y = 2x^2 + 3$$

$\_\_\_\_ = 2x^2 + 3$ Replace y with ____.

$1 - \_\_\_\_ = 2x^2 + 3 - \_\_\_\_$ Subtract ____ from each side.

$\_\_\_\_ = 2x^2$

$$\frac{-2}{\_\_\_\_} = \frac{2x^2}{\_\_\_\_}$$

$$-1 = x^2$$

$\_\_\_\_ = x$ Take the square root of each side.

There is no real solution because the square root of a negative number is not a real number.

FOLDABLES™

ORGANIZE IT

List the different methods of solving quadratic-linear systems under the tab for Lesson 13-6. Then describe the situations in which each method is preferred.

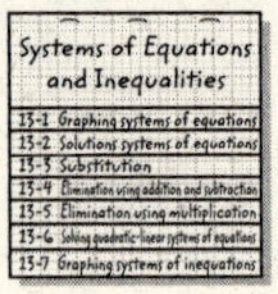

Your Turn **Use substitution to solve the system of equations.**

a. $y = -3x^2 - 4$
$y = -1$

b. $y = x^2 + 3$
$y = 7$

c. $y = x^2 - 6$
$y = x$

HOMEWORK ASSIGNMENT

Page(s):

Exercises:

13–7 Graphing Systems of Inequalities

WHAT YOU'LL LEARN

- Solve systems of inequalities by graphing.

BUILD YOUR VOCABULARY (page 271)

A set of two or more inequalities is a **system of inequalities**.

FOLDABLES™

ORGANIZE IT

Compare the solution of a linear system of equations with a linear system of inequalities under the tab for Lesson 13-7. Then give an example of each.

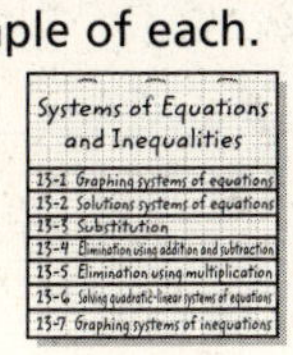

EXAMPLES

Solve each system of inequalities by graphing. If the system does not have a solution, write no solution.

1 $x \geq 3$

$y < x + 4$

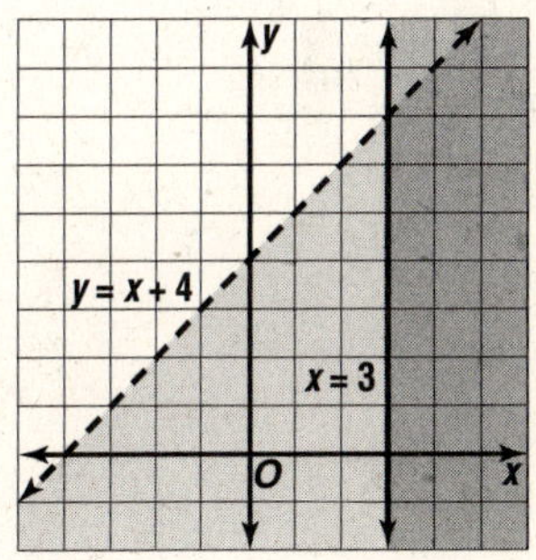

The solution is the ordered pairs in the ______ of the graphs of $x \geq 3$ and $y < x + 4$. Shade the region darkest on the graph. The graphs of $x \geq 3$ and $y < x + 4$ are the ______ of this region.

The graph of $y = x + 4$ is a ______ line and is ______ in the solution of the system. Choose a point and check the solution.

2 $x + y > 2$

$2y \leq -2x - 2$

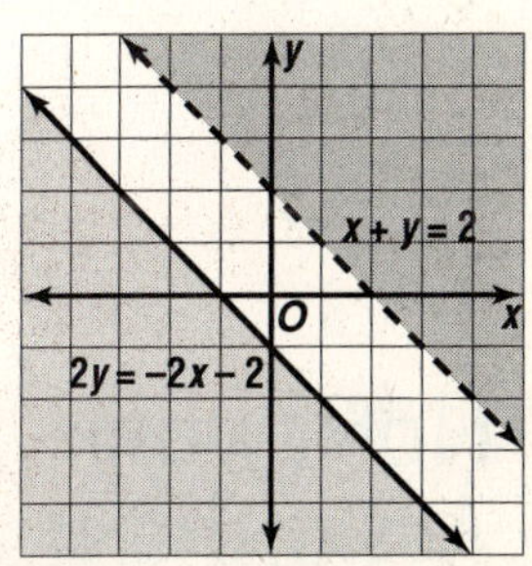

The graphs of $x + y = 2$ and $2y = -2x - 2$ are ______ lines. Check this by graphing or by comparing the ______.

Because the regions in the solution of $x + y > 2$ and $2y \leq -2x - 2$ have ______ in common, the system of inequalities has ______ solution.

Your Turn **Solve each system of inequalities by graphing. If the system does not have a solution, write no solution.**

a. $y > 2x - 1$
$3x + 2y \leq 6$

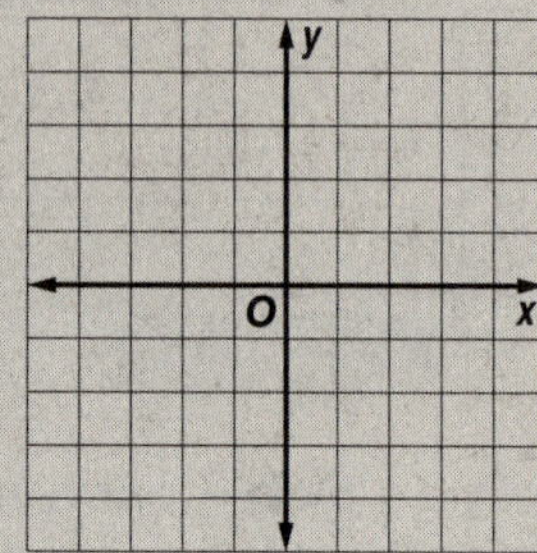

b. $y \leq 5$
$y < x - 2$

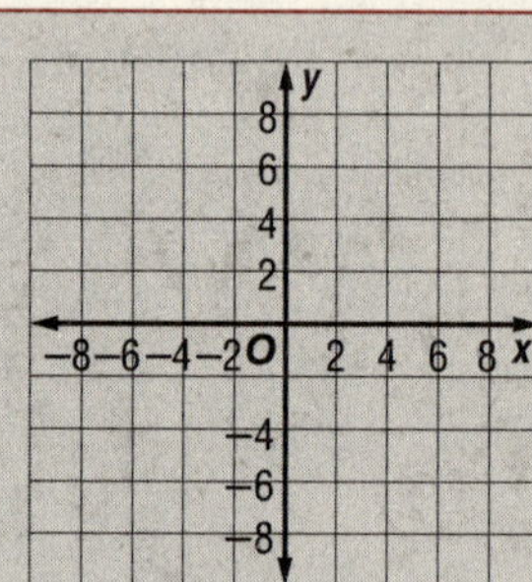

HOMEWORK ASSIGNMENT

Page(s):

Exercises:

CHAPTER 13

BRINGING IT ALL TOGETHER

STUDY GUIDE

FOLDABLES	VOCABULARY PUZZLEMAKER	BUILD YOUR VOCABULARY
Use your **Chapter 13 Foldable** to help you study for your chapter test.	To make a crossword puzzle, word search, or jumble puzzle of the vocabulary words in Chapter 13, go to: www.glencoe.com/sec/math/t_resources/free/index.php	You can use your completed **Vocabulary Builder** (pages 270–271) to help you solve the puzzle.

13-1

Graphing Systems of Equations

1. Solve the system of equations by graphing.

$y = 2x + 1$

$y = -3x + 6$

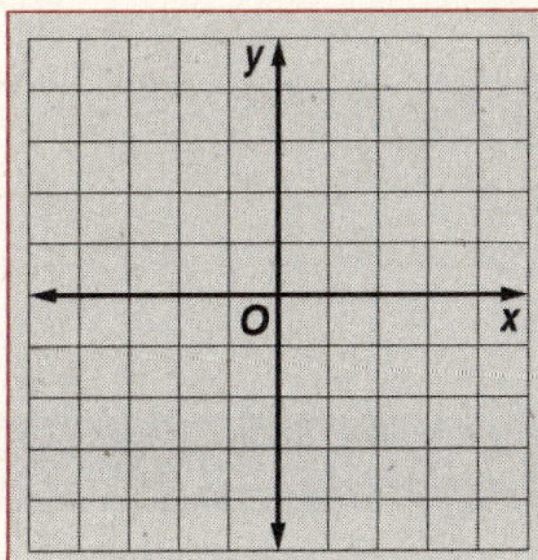

13-2

Solutions of Systems of Equations

Describe the solution of each system of equations.

2.

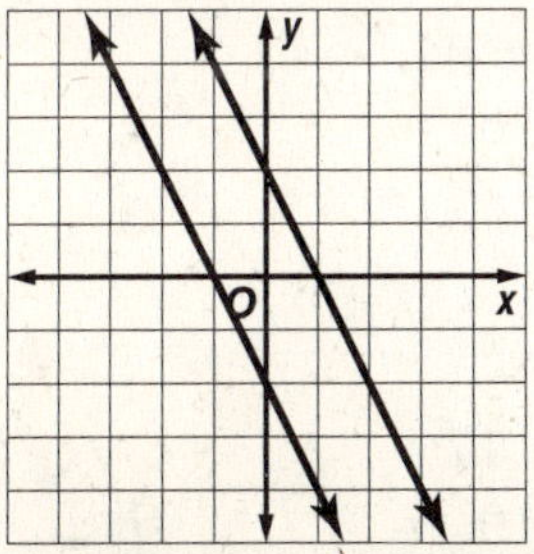

3.

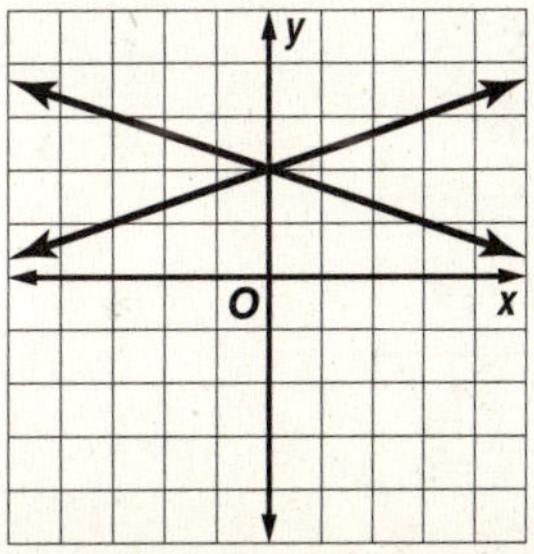

13-3

Substitution

4. Describe how you would use substitution to solve the system of equations. $y = -2x$
$x + 3y = 15$

5. Amy solved a system of equations and her result was $-8 = -8$. All of her work was correct. Describe the graph of the system. Explain.

6. Luis solved a system of equations and his result was $5 = -2$. All of his work was correct. Describe the graph of the system. Explain.

13-4

Elimination Using Addition and Subtraction

7. The sum of the digits of a two-digit number is 8. If the tens digit is 6 more than the units digit, what is the number?

Use elimination to solve each system of equations.

8. $3x + 5y = 15$
$-3x + 2y = 6$

9. $-x - 4y = -4$
$4x - 4y = 6$

10. $3x + 5y = 7$
$3x - 2y = 14$

11. $5x - 7y = 17$
$8x + 7y = 9$

13-5 Elimination Using Multiplication

Match each system with its solution.

12. $x + y = 4$
$-3x + 2y = 8$

13. $4x - 3y = -1$
$2x + y = 2$

14. $2x + y = 4$
$4x - y = 2$

15. $5x - 3y = -5$
$2x + 7y = -2$

a. $(4, -2)$
b. $(1, 2)$
c. $(0, 4)$
d. $\left(\frac{1}{2}, 1\right)$
e. $(-1, 0)$

13-6 Solving Quadratic-Linear Systems of Equations

Match each graph with the correct system of equations.

16.

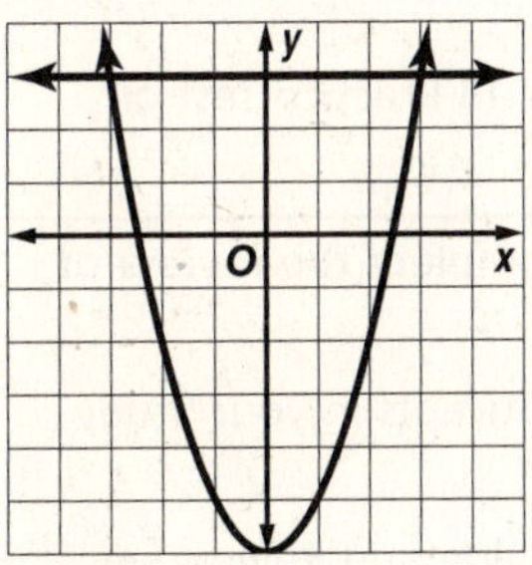

17.

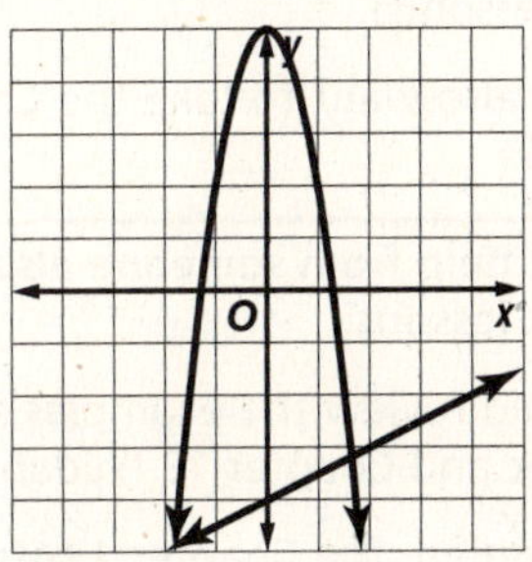

18.

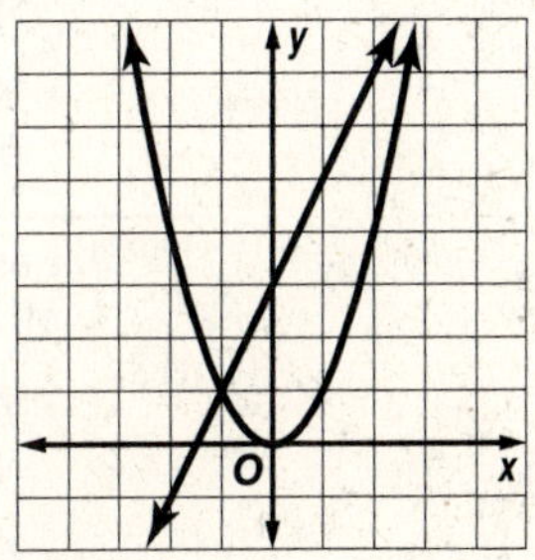

a.	**b.**	**c.**	**d.**
$y = x^2$	$y = x^2 - 6$	$y = \frac{1}{2}x - 4$	$y = 2x + 3$
$y = 2x + 3$	$y = 3$	$y = -3x^2 + 5$	$y = -x^2$

13-7 Graphing Systems of Inequalities

Write the inequality symbols that you need to get a system whose graph looks like the one shown. Use <, ≤, >, or ≥.

19.

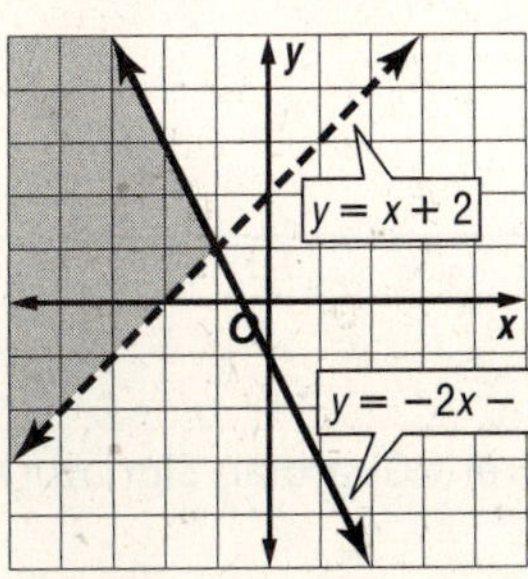

y ☐ $x + 2$

y ☐ $-2x - 1$

20.

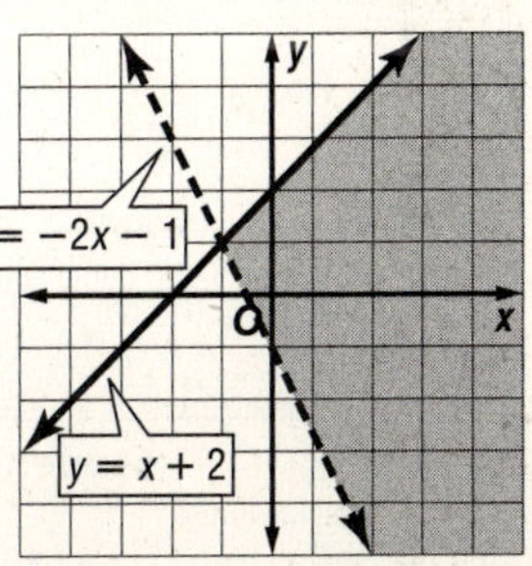

y ☐ $x + 2$

y ☐ $-2x - 1$

CHAPTER 13

Checklist

ARE YOU READY FOR THE CHAPTER TEST?

Math Online

Visit **algconcepts.com** to access your textbook, more examples, self-check quizzes, and practice tests to help you study the concepts in Chapter 13.

Check the one that applies. Suggestions to help you study are given with each item.

☐ **I completed the review of all or most lessons without using my notes or asking for help.**

- You are probably ready for the Chapter Test.
- You may want to take the Chapter 13 Practice Test on page 595 of your textbook as a final check.

☐ **I used my Foldable or Study Notebook to complete the review of all or most lessons.**

- You should complete the Chapter 13 Study Guide and Review on pages 592–594 of your textbook.
- If you are unsure of any concepts or skills, refer back to the specific lesson(s).
- You may also want to take the Chapter 13 Practice Test on page 595.

☐ **I asked for help from someone else to complete the review of all or most lessons.**

- You should review the examples and concepts in your Study Notebook and Chapter 13 Foldable.
- Then complete the Chapter 13 Study Guide and Review on pages 592–594 of your textbook.
- If you are unsure of any concepts or skills, refer back to the specific lesson(s).
- You may also want to take the Chapter 13 Practice Test on page 595.

Student Signature

Parent/Guardian Signature

Teacher Signature

CHAPTER 14

Radical Expressions

FOLDABLES™ Use the instructions below to make a Foldable to help you organize your notes as you study the chapter. You will see Foldable reminders in the margin of this Interactive Study Notebook to help you in taking notes.

Begin with a sheet of 11" × 17" paper.

STEP 1 **Fold**
Fold the short sides to meet in the middle.

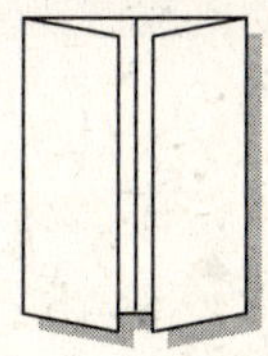

STEP 2 **Fold**
Fold the top to the bottom.

STEP 3 **Open**
Cut along the second fold to make four tabs.

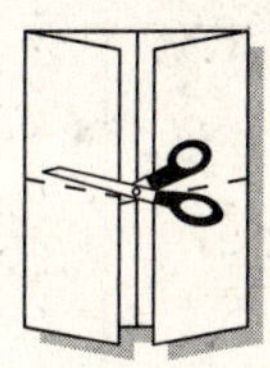

STEP 4 **Label**
Label each tab as shown.

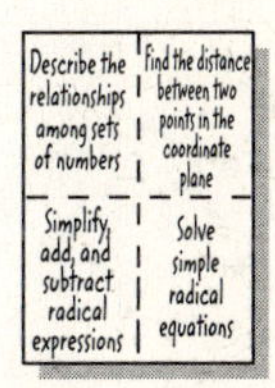

NOTE-TAKING TIP: As you study a lesson, write down questions you have, comments, reactions, and/or short summaries of the main ideas of the lesson that are highlighted or underlined.

CHAPTER 14

BUILD YOUR VOCABULARY

This is an alphabetical list of new vocabulary terms you will learn in Chapter 14. As you complete the study notes for the chapter, you will see Build Your Vocabulary reminders to complete each term's definition or description on these pages. Remember to add the textbook page number in the second column for reference when you study.

Vocabulary Term	Found on Page	Definition	Description or Example
conjugates [CON-ja-guts]			
Distance Formula			
radical equations			
radicand [RA-di-KAND]			
rationalizing the denominator [RA-shun-ul-eyes-ing]			

14–1 The Real Numbers

EXAMPLES

WHAT YOU'LL LEARN

- Describe the relationships among sets of numbers.

FOLDABLES

ORGANIZE IT

Explain the similarities and differences between the sets of natural numbers, whole numbers, integers, rational numbers, irrational numbers, and real numbers under the tab for "Describe the relationships among sets of numbers."

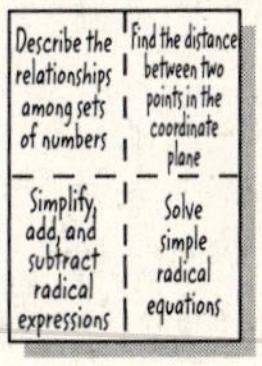

Name the set or sets of numbers to which each real number belongs.

1 **10**

This number is a natural number, a ______ number, an ______, and a ______ number.

2 $-\sqrt{100}$

Since $-\sqrt{100} =$ ______, this number is an ______ and a rational number.

3 5^2

Since $5^2 =$ ______, this number is a natural number, a ______ number, an ______, and a rational number.

4 **−0.666 . . .**

This repeating decimal is a ______ number since it is equivalent to $-\frac{2}{3}$.

5 $\sqrt{17}$

$\sqrt{17} = 4.123105626\ldots$ It is not the square root of a perfect square. So, it is an ______ number.

Your Turn **Name the set or sets of numbers to which each real number belongs.**

a. $\sqrt{36}$ ______

b. $-\frac{18}{3}$ ______

c. $-\sqrt{2}$

d. -2^3

e. $0.\overline{5}$

EXAMPLES

Find an approximation, to the nearest tenth, for each square root. Then graph the square root on a number line.

KEY CONCEPT

Completeness Property for Points on the Number Line Each real number corresponds to exactly one point on the number line. Each point on the number line corresponds to exactly one real number.

6 $\sqrt{10}$

Enter: 2nd [$\sqrt{\ }$] ENTER 3.16227766

An approximate value for $\sqrt{10}$ is ____.

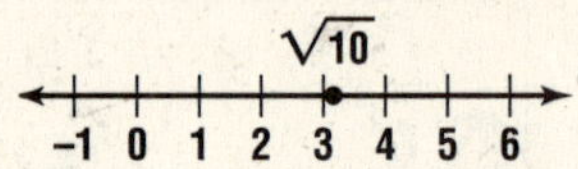

7 $-\sqrt{8}$

Enter: (–) 2nd [$\sqrt{\ }$] 8 ENTER -2.828427125

An approximate value for $-\sqrt{8}$ is $-$____.

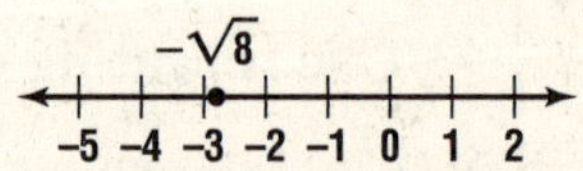

Your Turn **Find an approximation, to the nearest tenth, for each square root. Then graph the square root on a number line.**

a. $\sqrt{7}$

b. $-\sqrt{19}$

EXAMPLES

Determine whether each number is rational or irrational. If it is irrational, find two consecutive integers between which its graph lies on the number line.

REMEMBER IT

To find the two consecutive integers between which an irrational square root lies, make a list of perfect square numbers: 1, 4, 9, 16, 25, 36, 49, 64, 81, . . . Then find the two numbers between which the number under the square root symbol lies.

8 $-\sqrt{59}$

_____ $< -\sqrt{59} <$ _____

_____ $< -\sqrt{59} <$ _____

59 is not a _____. So, its square root is _____. The graph of $-\sqrt{59}$ lies between _____ and _____.

9 $\sqrt{121}$

Since $\sqrt{121} =$ _____, it is a _____.

Your Turn **Determine whether each number is rational or irrational. If it is irrational, find two consecutive integers between which its graph lies on the number line.**

a. $-\sqrt{64}$

b. $\sqrt{23}$

HOMEWORK ASSIGNMENT

Page(s):

Exercises:

14–2 The Distance Formula

WHAT YOU'LL LEARN

- Find the distance between two points in the coordinate plane.

BUILD YOUR VOCABULARY (page 298)

The **Distance Formula** is a formula derived from the ______ Theorem to find the distance between two ______ on the coordinate plane.

EXAMPLES

KEY CONCEPT

The Distance Formula The distance d between any two points with coordinates (x_1, y_1) and (x_2, y_2) is given by $d = \sqrt{(x_2 - x_1)^2 + (y_2 - y_1)^2}$.

1 **Find the distance between points $E(-3, 8)$ and $F(5, 1)$.**

$d = \sqrt{(x_2 - x_1)^2 + (y_2 - y_1)^2}$ Distance Formula

$= \sqrt{(\_\_\_\_\_\_ - \_\_\_\_\_\_)^2 + (\_\_\_\_\_\_ - \_\_\_\_\_\_)^2}$ $(x_1, y_1) = (-3, 8)$, and $(x_2, y_2) = (5, 1)$

$= \sqrt{\_\_\_\_\_\_^2 + (\_\_\_\_\_\_)^2}$

$= \sqrt{\_\_\_\_\_\_ + \_\_\_\_\_\_}$

$= \_\_\_\_\_\_$

The distance between points $E(-3, 8)$ and $F(5, 1)$ is ______ or about ______ units.

2 **A coordinate system is superimposed over a map of Washington, D.C. Two taxis leave the intersection of 15th Street and G Street. One taxi travels 7 blocks north, and the other taxi travels 11 blocks east. How far apart are the taxis when they stop?**

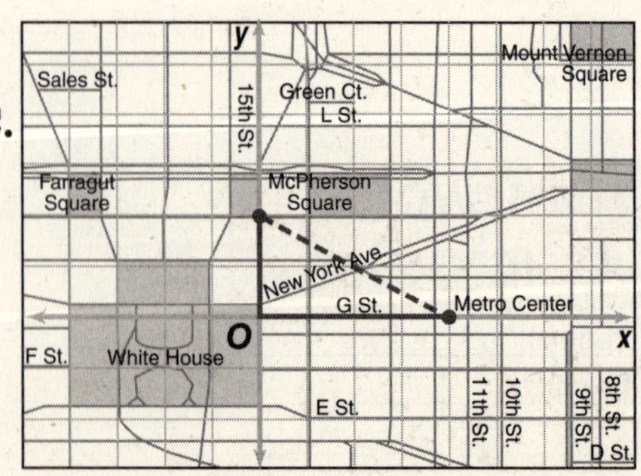

Let the first taxi's location be represented by (0, 7) and the other taxi's location by (11, 0). So, $x_1 =$ ____,

$y_1 =$ ____, $x_2 =$ ____, and $y_2 =$ ____.

$d = \sqrt{(x_2 - x_1)^2 + (y_2 - y_1)^2}$ Distance Formula

$= \sqrt{(\_\_\_\_)^2 + (\_\_\_\_)^2}$ $(x_1, y_1) = (0, 7)$, and $(x_2, y_2) = (11, 0)$

$= \sqrt{\_\_\_\_^2 + (\_\_\_\_)^2}$

$= \sqrt{\_\_\_\_ + \_\_\_\_}$

$=$ ____ or about ____ units

The taxis are about ____ blocks apart when they stop.

FOLDABLES™

ORGANIZE IT

Write the distance formula and two examples to help you remember how to use it under the tab for "Find the distance between two points in the coordinate plane."

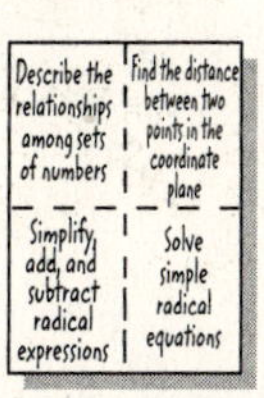

Your Turn

a. Find the distance between points $C(2, 7)$ and $D(6, -4)$.

b. A coordinate system is superimposed over a map of Jake's neighborhood. Jake leaves his school and travels 3 blocks west to a friend's house. Then he turns and travels 5 blocks south to his home. How far is Jake's school from his house?

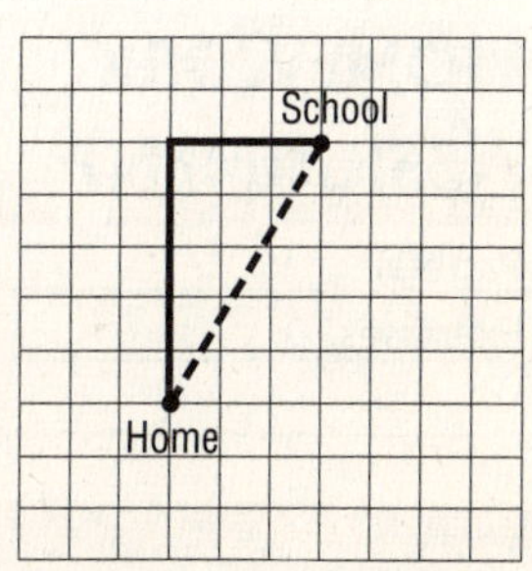

EXAMPLE

3 Find the value of a if $M(a, 6)$ and $N(2, 3)$ are $\sqrt{18}$ units apart.

$$d = \sqrt{(x_2 - x_1)^2 + (y_2 - y_1)^2}$$ Distance Formula

$$\square = \sqrt{(\square)^2 + (\square)^2}$$ $(x_1, y_1) = (a, 6)$, and $(x_2, y_2) = (2, 3)$

$$\square = \sqrt{(\square) + (\square)^2}$$

$$\sqrt{18} = \sqrt{(4 - 4a + a^2) + 9}$$ $(2 - a)^2 = 4 - 4a + a^2$ and $(-3)^2 = 9$

$$\sqrt{18} = \sqrt{a^2 - 4a + 13}$$

$$(\sqrt{18})^2 = (\sqrt{a^2 - 4a + 13})^2$$ Square each side.

$$\square = a^2 - 4a + 13$$

$$0 = \square$$ Subtract.

$$0 = (a + 1)(a - 5)$$ Factor.

$$\square = 0 \text{ or } \square = 0$$ Zero Product Property

$$a = \square \qquad a = \square.$$

The value of a is ☐ or ☐.

REVIEW IT

The Zero Product Property says that for all numbers a and b, if $ab = 0$, then $b = 0$, $b = 0$, or both a and b equal 0. *(Lesson 11-4)*

Your Turn Find the value of a if $P(-3, 5)$ and $Q(2, a)$ are $\sqrt{194}$ units apart.

HOMEWORK ASSIGNMENT

Page(s):

Exercises:

14–3 Simplifying Radical Expressions

BUILD YOUR VOCABULARY (page 298)

The **radicand** is the number under the ________ sign.

Rationalizing the denominator is a method used to remove ________ from the ________ of a fraction.

Conjugates are two expressions of the form $a\sqrt{b} + c\sqrt{d}$ and $a\sqrt{b} - c\sqrt{d}$.

WHAT YOU'LL LEARN

- Simplify radical expressions.

REVIEW IT

The Product Property of Square Roots says that the square root of a product is equal to the product of each square root. In symbols, $\sqrt{ab} = \sqrt{a} \cdot \sqrt{b}$ if $a \geq 0$ and $b \geq 0$. *(Lesson 8-5)*

EXAMPLE

1 Simplify $\sqrt{48}$.

$\sqrt{48} = \sqrt{\quad\quad}$	Prime factorization
$= \sqrt{16 \cdot 3}$	$2 \times 2 \times 2 \times 2 = 16$
$= \sqrt{16} \cdot$ ____	Product Property of Square Roots
$=$ ____	Simplify $\sqrt{16}$.

EXAMPLE

2 Simplify $\sqrt{15} \cdot \sqrt{75}$.

$\sqrt{15} \cdot \sqrt{75} = \sqrt{3 \cdot 5} \cdot \sqrt{3 \cdot 5 \cdot 5}$	Prime factorization
$= \sqrt{3} \cdot \sqrt{5} \cdot \sqrt{3} \cdot \sqrt{5} \cdot \sqrt{5}$	Product Property of Square Roots
$=$ ____ $\cdot \sqrt{5} \cdot \sqrt{5} \cdot \sqrt{5}$	Commutative Property
$= \sqrt{3^2} \cdot \sqrt{5^2} \cdot$ ____	$\sqrt{3} \cdot \sqrt{3} = \sqrt{3^2}$ and $\sqrt{5} \cdot \sqrt{5} = \sqrt{5^2}$
$= 3 \cdot 5 \cdot$ ____	
$=$ ____	Simplify.

Your Turn **Simplify each expression.**

a. $\sqrt{72}$ ______

b. $\sqrt{6} \cdot \sqrt{12}$ ______

EXAMPLE

3 **Simplify** $\dfrac{\sqrt{300}}{\sqrt{15}}$.

$\dfrac{\sqrt{300}}{\sqrt{15}} = \sqrt{\dfrac{300}{15}}$ Quotient Property of Square Roots

$= \sqrt{20}$ Divide 300 by 15.

$= \sqrt{2 \cdot 2 \cdot 5}$ Prime factorization

$= \sqrt{\boxed{}^{2} \cdot 5}$ $2 \cdot 2 = 2^2$

$= \sqrt{2^2} \cdot \boxed{}$ Product Property of Square Roots

$= \boxed{}$ Simplify.

REVIEW IT

The Quotient Property of Square Roots says that the square root of a quotient is equal to the quotient of each square root. In symbols, $\sqrt{\dfrac{a}{b}} = \dfrac{\sqrt{a}}{\sqrt{b}}$ if $a \geq 0$ and $b > 0$. *(Lesson 8-5)*

EXAMPLE

4 **Simplify** $\dfrac{\sqrt{6}}{\sqrt{30}}$.

$\dfrac{\sqrt{6}}{\sqrt{30}} = \dfrac{\sqrt{6}}{\sqrt{30}} \cdot \dfrac{\sqrt{30}}{\sqrt{30}}$ $\dfrac{\sqrt{30}}{\sqrt{30}} = 1$

$= \dfrac{\sqrt{6 \cdot 30}}{\sqrt{30 \cdot 30}}$

$= \dfrac{\sqrt{\boxed{}}}{\sqrt{900}}$ Simplify.

$= \dfrac{\sqrt{9 \cdot 4 \cdot 5}}{30}$

$= \dfrac{\boxed{} \cdot \boxed{} \cdot \boxed{}}{30}$ Product Property of Square Roots

$= \dfrac{3 \cdot 2 \cdot \sqrt{5}}{30}$

$= \boxed{}$ Simplify.

KEY CONCEPT

Simplified Form for Radicals A radical expression is in simplest form when:

- No radicands have perfect square factors other than 1.
- No radicands contain fractions.
- No radicals appear in the denominator of a fraction.

Your Turn **Simplify each expression.**

a. $\dfrac{\sqrt{16}}{\sqrt{48}}$

b. $\dfrac{\sqrt{12}}{\sqrt{72}}$

EXAMPLES

5 **Simplify $\dfrac{5}{4 + \sqrt{3}}$.**

To rationalize the denominator, multiply both the numerator and denominator by the conjugate of $4 + \sqrt{3}$.

$\dfrac{5}{4+\sqrt{3}} = \dfrac{5}{4+\sqrt{3}} \cdot \dfrac{4-\sqrt{3}}{4-\sqrt{3}}$ — $\dfrac{4-\sqrt{3}}{4-\sqrt{3}} = 1$

$= \dfrac{5(4) - 5\sqrt{3}}{\square^2 - \square}$ — Distributive Property; $(a - b)(a + b) = a^2 - b^2$

$= \dfrac{\square}{16 - 3}$ — Simplify.

$= \dfrac{20 - 5\sqrt{3}}{13}$

6 **Simplify $\sqrt{48y^3z^6}$.**

$\sqrt{48y^3z^6}$

$= \sqrt{2 \cdot 2 \cdot 2 \cdot 2 \cdot 3 \cdot y^3 \cdot z^6}$ — Prime factorization

$= \sqrt{16 \cdot 3 \cdot \square \cdot y^2 \cdot z^6}$ — $2 \times 2 \times 2 \times 2 = 16$

$= \square \cdot \sqrt{3} \cdot \sqrt{y} \cdot \sqrt{y^2} \cdot \sqrt{z^6}$ — Product Property of Square Roots

$= 4 \cdot \sqrt{3} \cdot \sqrt{y} \cdot \square \cdot z^3$ — Simplify.

$= 4|y|z^3\sqrt{3y}$ — The absolute value of y ensures a nonnegative result.

HOMEWORK ASSIGNMENT

Page(s):

Exercises:

Your Turn **Simplify each expression.**

a. $\dfrac{7}{2 - \sqrt{5}}$

b. $\sqrt{108a^4b^3}$

14–4 Adding and Subtracting Radical Expressions

EXAMPLES

WHAT YOU'LL LEARN

- Add and subtract radical expressions.

Simplify each expression.

1 $\mathbf{12\sqrt{10} + 15\sqrt{10}}$

$12\sqrt{10} + 15\sqrt{10}$

$= (\square)\sqrt{10}$ Distributive Property

$= \square$

2 $\mathbf{3\sqrt{2} - 9\sqrt{2}}$

$3\sqrt{2} - 9\sqrt{2} = (\square)\sqrt{2}$ Distributive Property

$= \square$

3 **The lengths of the three sides of a triangle are $\mathbf{10\sqrt{3}}$ centimeters, $\mathbf{5\sqrt{5}}$ centimeters, and $\mathbf{9\sqrt{3}}$ centimeters. Find the exact perimeter of the triangle.**

REMEMBER IT

To add or subtract radicals, the radicands must be the same.

$P = \square + \square + \square$ Like terms: $10\sqrt{3}$ and $9\sqrt{3}$

$= 10\sqrt{3} + 9\sqrt{3} + 5\sqrt{5}$ Commutative Property

$= (10 + 9)\sqrt{3} + \square$ Distributive Property

$= \square + \square$

The exact perimeter of the triangle is $\square + \square$ centimeters.

Your Turn **Simplify each expression.**

a. $-8\sqrt{5} + 12\sqrt{5}$

b. $2\sqrt{11} + 4\sqrt{11} - 9\sqrt{11}$

c. The lengths of the sides of a quadrilateral are $8\sqrt{2}$ inches, $3\sqrt{7}$ inches, $5\sqrt{7}$ inches, and $6\sqrt{2}$ inches. Find the exact perimeter of the quadrilateral.

FOLDABLES™

ORGANIZE IT

Write a note to explain the process for simplifying, adding, and subtracting radical expressions under the tab for "Simplify, add and subtract radical expressions." Then show one example of each.

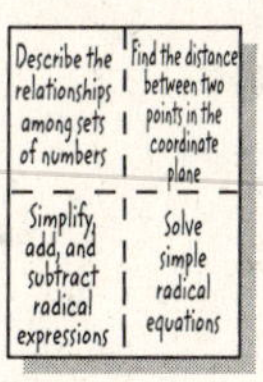

EXAMPLE

4 **Simplify $2\sqrt{48} - 3\sqrt{75}$.**

$2\sqrt{48} - 3\sqrt{75}$

$= 2\sqrt{\square^4 \cdot 3} - 3\sqrt{\square^2 \cdot 3}$ Prime factorization

$= 2 \cdot \left(\square \cdot \sqrt{3}\right) - 3\left(\square \cdot \sqrt{3}\right)$ Simplify.

$= 2 \cdot 4 \cdot \sqrt{3} - 15\sqrt{3}$

$= \square - 15\sqrt{3}$

$= (8 - 15)\sqrt{3}$ Distributive Property

$= \square$

Your Turn Simplify $3\sqrt{72} + 5\sqrt{18}$.

HOMEWORK ASSIGNMENT

Page(s):

Exercises:

14–5 Solving Radical Equations

Build Your Vocabulary (page 298)

A **radical equation** is an equation that contains a ______ expression.

What You'll Learn

- Solve simple radical equations in which only one radical contains a variable.

Foldables™

Organize It

Explain how to eliminate the radical from an equation under the tab for "Solve simple radical equations." Then show two examples.

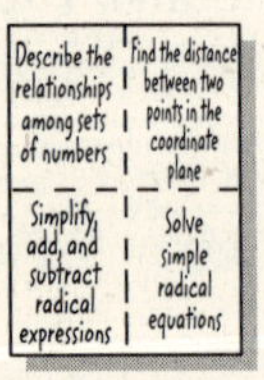

Examples

Solve each equation. Check your solution.

1 $3 + \sqrt{x} = 12$

$3 + \sqrt{x} = 12$

$3 + \sqrt{x} - \square = 12 - \square$ Subtract $\square$ from each side.

$\sqrt{x} = 9$

$(\sqrt{x})^2 = \square^2$ Square each side.

$x = \square$

Check: $3 + \sqrt{x} = 12$

$3 + \sqrt{\square} \stackrel{?}{=} 12$ Replace x with $\square$.

$3 + \square \stackrel{?}{=} 12$

$12 = 12$ ✓

2 $-6 + \sqrt{y - 2} = 1$

$-6 + \sqrt{y - 2} = 1$

$-6 + \sqrt{y - 2} + \square = 1 + \square$ Add $\square$ to each side.

$\sqrt{y - 2} = \square$

$(\sqrt{y - 2})^2 = \square^2$ Square each side.

$y - 2 = 49$

$y - 2 + 2 = 49 + 2$ Add 2 to each side.

$y = \square$ Check this result.

REMEMBER IT When solving a radical equation, always isolate the radical *before* squaring each side.

Your Turn **Solve each equation. Check your solution.**

a. $-2 + \sqrt{x} = 9$

b. $\sqrt{z + 5} - 1 = 8$

EXAMPLES

Solve each equation. Check your solution.

3 $\sqrt{3m + 4} = m$

$\sqrt{3m + 4} = m$

$(\sqrt{3m + 4})^2 = m^2$ Square each side.

$\square = m^2$

$3m - 3m + 4 - 4 = m^2 - 3m - 4$ Subtract $3m$ and 4 from each side.

$0 = \square$

$0 = (m - \square)(m + \square)$ Factor.

$m - \square = 0$ or $m + \square = 0$ Zero Product Property

$m = \square$ $m = \square$

Check: $\sqrt{3m + 4} = m$ $\sqrt{3m + 4} = m$

$\sqrt{3(\square) + 4} \stackrel{?}{=} 4$ $\sqrt{3(\square) + 4} \stackrel{?}{=} -1$

$\sqrt{\square + 4} \stackrel{?}{=} 4$ $\sqrt{\square + 4} \stackrel{?}{=} -1$

$\sqrt{\square} \stackrel{?}{=} 4$ $\sqrt{\square} \stackrel{?}{=} -1$

$4 = 4$ $1 \neq -1$

Since $\square$ does not satisfy the original equation, $\square$ is the only solution.

WRITE IT

Why is it important to check your solutions when solving radical equations?

______

______

______

______

______

4 $\sqrt{n+5} = n + 5$

$$\sqrt{n+5} = n + 5$$

$$(\sqrt{n+5})^2 = (n+5)^2 \quad \text{Square each side.}$$

$$n + 5 = n^2 + 10n + 25$$

$$n - n + 5 - 5 = n^2 + 10n + 25 - n - 5 \quad \text{Subtract.}$$

$$0 = n^2 + \square + 20$$

$$0 = (n + \square)(n + \square) \quad \text{Factor.}$$

$$n + \square = 0 \text{ or } n + \square = 0 \quad \text{Zero Product Property}$$

$$n = \square \qquad n = \square$$

Check: $\sqrt{n+5} = n + 5$ | $\sqrt{n+5} = n + 5$

$$\sqrt{(\square) + 5} \stackrel{?}{=} (\square) + 5 \qquad \sqrt{(\square) + 5} \stackrel{?}{=} (\square) + 5$$

$$\sqrt{0} \stackrel{?}{=} 0 \qquad \sqrt{1} \stackrel{?}{=} 1$$

$$0 = 0 ✓ \qquad 1 = 1 ✓$$

Since both ☐ and ☐ satisfy the original equation, they are both solutions.

Your Turn **Solve each equation. Check your solution.**

a. $y = \sqrt{3y + 10}$

b. $\sqrt{x+3} - x = 3$

HOMEWORK ASSIGNMENT

Page(s):

Exercises:

CHAPTER 14

BRINGING IT ALL TOGETHER

STUDY GUIDE

FOLDABLES™	VOCABULARY PUZZLEMAKER	BUILD YOUR VOCABULARY
Use your **Chapter 14 Foldable** to help you study for your chapter test.	To make a crossword puzzle, word search, or jumble puzzle of the vocabulary words in Chapter 14, go to: www.glencoe.com/sec/math/t_resources/free/index.php	You can use your completed **Vocabulary Builder** (page 298) to help you solve the puzzle.

14-1 The Real Numbers

For each of the following, choose the letter of each set of numbers to which each real number belongs. Each real number may belong to more than one set of numbers.

1. 3.6

2. −5

3. $\sqrt{41}$

4. 0

5. 120

a. Natural numbers
b. Whole numbers
c. Integers
d. Rational numbers
e. Irrational numbers

14-2 The Distance Formula

6. Suppose you want to use the Distance Formula to find the distance between $M(3, 7)$ and $N(9, -2)$. Use $(x_1, y_1) = (3, 7)$ and $(x_2, y_2) = (9, -2)$. Complete the equation by writing the correct numbers in the boxes.

$$d = \sqrt{(\square - \square)^2 + (\square - \square)^2}$$

7. What is the value of b if $S(7, b)$ and $T(13, 1)$ are 10 units apart?

14-3

Simplifying Radical Expressions

8. Of $5\sqrt{3}$, $5\sqrt{6}$, and $3\sqrt{3}$, which two radical expressions have the same radicand?

Simplify each expression. Leave in radical form.

9. $\sqrt{32}$

10. $\sqrt{6} \cdot \sqrt{18}$

11. $\dfrac{2}{1 - \sqrt{7}}$

12. a. How can you tell that the radical expression $\sqrt{28x^2y^4}$ is *not* in simplest form?

b. To simplify $\sqrt{28x^2y^4}$, you first find the ________ of $28x^2y^4$. Then apply the ________ Property of Square Roots. In this case, $4 \cdot 7 \cdot x^2 \cdot y^4$ is equal to the product ________. Simplify again to get a final answer of $2\,|x|\,y^2\sqrt{7}$.

13. What method would you use to simplify $\dfrac{\sqrt{12t}}{\sqrt{15}}$?

14-4 Adding and Subtracting Radical Expressions

14. Radical expressions can be added or subtracted if they have the same ________.

15. Indicate whether the following expressions are in simplest form. Explain your answer.

a. $6\sqrt{3} - \sqrt{12}$

b. $12\sqrt{6} + 7\sqrt{10}$

c. $4\sqrt{15} - 3\sqrt{60}$

d. $3\sqrt{20} + 5\sqrt{30}$

14-5 Solving Radical Equations

16. Provide the reason for each step in the solution of the given radical equation.

$\sqrt{5x - 1} - 4 = x - 3$	Original equation
$\sqrt{5x - 1} = x + 1$	
$(\sqrt{5x - 1})^2 = (x + 1)^2$	
$5x - 1 = x^2 + 2x + 1$	
$0 = x^2 - 3x + 2$	
$0 = (x - 1)(x - 2)$	
$x - 1 = 0$ or $x - 2 = 0$	
$x = 1$ or $x = 2$	

CHAPTER 14

Checklist

ARE YOU READY FOR THE CHAPTER TEST?

Math Online

Visit **algconcepts.com** to access your textbook, more examples, self-check quizzes, and practice tests to help you study the concepts in Chapter 14.

Check the one that applies. Suggestions to help you study are given with each item.

☐ **I completed the review of all or most lessons without using my notes or asking for help.**

- You are probably ready for the Chapter Test.
- You may want take the Chapter 14 Practice Test on page 633 of your textbook as a final check.

☐ **I used my Foldable or Study Notebook to complete the review of all or most lessons.**

- You should complete the Chapter 14 Study Guide and Review on pages 630–632 of your textbook.
- If you are unsure of any concepts or skills, refer back to the specific lesson(s).
- You may also want to take the Chapter 14 Practice Test on page 633.

☐ **I asked for help from someone else to complete the review of all or most lessons.**

- You should review the examples and concepts in your Study Notebook and Chapter 14 Foldable.
- Then complete the Chapter 14 Study Guide and Review on pages 630–632 of your textbook.
- If you are unsure of any concepts or skills, refer back to the specific lesson(s).
- You may also want to take the Chapter 14 Practice Test on page 633.

Student Signature

Parent/Guardian Signature

Teacher Signature

CHAPTER 15

Rational Expressions and Equations

FOLDABLES™ Use the instructions below to make a Foldable to help you organize your notes as you study the chapter. You will see Foldable reminders in the margin of this Interactive Study Notebook to help you in taking notes.

Begin with a sheet of notebook paper.

STEP 1 **Fold**
Fold lengthwise to the holes.

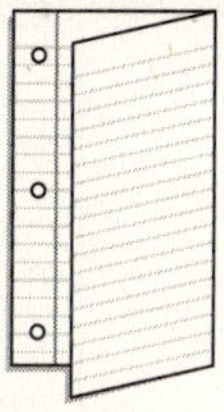

STEP 2 **Cut**
Cut along the top line and then cut ten tabs.

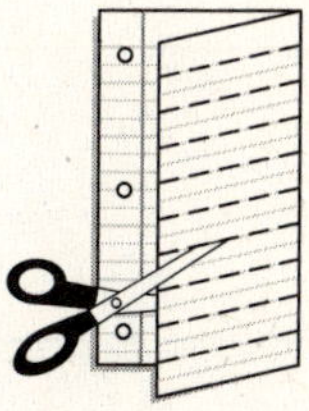

STEP 3 **Label**
Lable the tabs using the vocabulary words in the chapter.

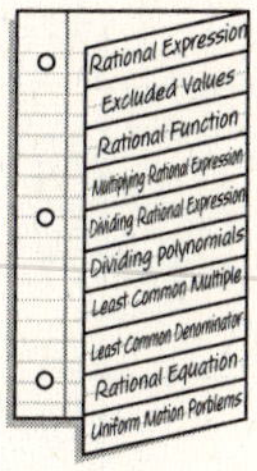

NOTE-TAKING TIP: When taking notes, always write definitions and examples of each of the terms learned.

Chapter 15

CHAPTER 15

BUILD YOUR VOCABULARY

This is an alphabetical list of new vocabulary terms you will learn in Chapter 15. As you complete the study notes for the chapter, you will see Build Your Vocabulary reminders to complete each term's definition or description on these pages. Remember to add the textbook page number in the second column for reference when you study.

Vocabulary Term	Found on Page	Definition	Description or Example
excluded value			
least common denominator (LCD)			
least common multiple (LCM)			
rational equation [RA-shun-ul]			
rational expression			
rational function			
uniform motion problem			
work problems			

15–1 Simplify Rational Expressions

What You'll Learn

- Simplify rational expressions.

Build Your Vocabulary (page 318)

An **excluded value** is any value assigned to a ______ that results in a denominator of ______.

Key Concept

Rational Expression
A rational expression is an algebraic fraction whose numerator and denominator are polynomials.

EXAMPLES

Find the excluded value(s) for each rational expression.

1 $\frac{20}{(x+3)x}$

Exclude the values for which $(x + 3)x =$ ______.

$(x + 3)x =$ ______

______ $= 0$ or ______ $= 0$ ______ Product Property

So, x cannot equal ______ or ______.

2 $\frac{10y}{y^2 - 4}$

Exclude the values for which $y^2 - 4 =$ ______.

$y^2 - 4 =$ ______

______ $= 0$ Factor ______.

$y =$ ______ or $y =$ ______ Zero Product Property

So, y cannot equal ______ or ______.

Your Turn **Find the excluded value(s) for each rational expression.**

a. $\frac{3}{m(7 - m)}$

b. $\frac{-y}{y^2 - 16}$

EXAMPLES

REMEMBER IT

Division by zero is undefined.

Simplify each rational expression.

3 $\dfrac{20x^5y}{25x^2y^3}$

$\dfrac{20x^5y}{25x^2y^3} = \dfrac{2 \cdot 2 \cdot 5 \cdot x \cdot x \cdot x \cdot x \cdot x \cdot y}{\underline{\qquad\qquad}}$

Note that $x \neq 0$ and $y \neq 0$.

$= \dfrac{2 \cdot 2 \cdot \overset{1}{\cancel{5}} \cdot \overset{1}{\cancel{x}} \cdot \overset{1}{\cancel{x}} \cdot x \cdot x \cdot x \cdot \overset{1}{\cancel{y}}}{\underline{\qquad\qquad}}$ or $\dfrac{4x^3}{\underline{\qquad}}$

The GCF is $5x^2y$.

4 $\dfrac{4x + 12}{7x + 21}$

$\dfrac{4x + 12}{7x + 21} = \dfrac{\underline{\qquad}(x + 3)}{\underline{\qquad}(x + 3)}$

Factor $4x + 12$ and $7x + 21$.

$= \dfrac{\underline{\qquad}\overset{1}{\cancel{(x + 3)}}}{\underline{\qquad}\underset{1}{\cancel{(x + 3)}}}$ or $\dfrac{\underline{\qquad}}{\underline{\qquad}}$

The GCF is $\underline{\qquad}$.

FOLDABLES™

ORGANIZE IT

Write an example of a rational expression under the tab for Rational Expression. Then find all the excluded values for the expression and write those under the tab for Excluded Values. Write the rational expression as a function and draw its graph on the tab for Rational Function.

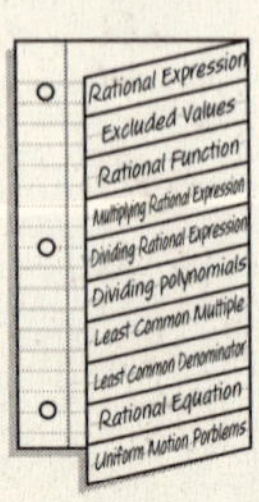

Your Turn **Simplify each rational expression.**

a. $\dfrac{18a^2b}{21a^5b^4}$

b. $\dfrac{3x - 12}{5x - 20}$

EXAMPLES

Simplify each rational expression.

REVIEW IT

Explain how to factor a trinomial. How do you find the correct combination of inner and outer terms? *(Lesson 10-3)*

5 $\dfrac{a^2 - 25}{a^2 - 2a - 15}$

$\dfrac{a^2 - 25}{a^2 - 2a - 15} = \dfrac{(a + \square)(a - \square)}{(a + \square)(a - \square)}$ Factor $a^2 - 25$ and $a^2 - 2a - 15$.

$= \dfrac{(a + 5)(\square)}{(a + 3)(\square)}$ with 1 above and 1 below, or $\dfrac{(\square)}{a + 3}$

The GCF is $\square$.

6 $\dfrac{6 - 2x}{x^2 - x - 12}$

$\dfrac{6 - 2x}{x^2 + x - 12} = \dfrac{2(3 - x)}{(x + \square)(x - \square)}$ Factor $6 - 2x$ and $x^2 + x - 12$.

$= \dfrac{2(-1)(x - 3)}{(x + 4)(\square)}$ Factor -1 from $\square$.

$= \dfrac{2(-1)(\square)}{(x + 4)(\square)}$ with 1 above and 1 below, or $= \dfrac{\square}{x + 4}$ The GCF is $x - 3$.

Your Turn **Simplify each rational expression.**

a. $\dfrac{15 - 5x}{x^2 + 2x - 15}$

b. $\dfrac{b^2 + 12b + 27}{b^2 - 9}$

HOMEWORK ASSIGNMENT

Page(s):

Exercises:

15–2 Multiplying and Dividing Rational Expressions

What You'll Learn

- Multiply and divide rational expressions.

EXAMPLES

Find each product.

1 $\frac{10a^2}{8ab} \cdot \frac{4ab}{10a^5}$

$$\frac{10a^2}{8ab} \cdot \frac{4ab}{10a^5} = \frac{\overset{1\ 1}{\cancel{10}\cancel{a^2}}}{\underset{2}{\cancel{8ab}}} \cdot \frac{\overset{1}{\cancel{4ab}}}{\underset{1\ a^3}{\cancel{10}\cancel{a^5}}}$$ Simplify.

$$= \frac{\square}{\square}$$ Multiply.

2 $\frac{z+3}{z-5} \cdot \frac{3z^2-15z}{4z+12}$

$$\frac{z+3}{z-5} \cdot \frac{3z^2-15z}{4z+12} = \frac{z+3}{z-5} \cdot \frac{\square(z-5)}{\square(z+3)}$$ Factor $3z^2 - 15z$ and $4z + 12$.

$$= \frac{\overset{1}{\square}}{\underset{1}{\cancel{z-5}}} \cdot \frac{\overset{1}{3z(\cancel{z-5})}}{\underset{1}{4\square}}$$ $(z + 3)$ and $(z - 5)$ are common factors.

$$= \frac{\square}{4}$$

3 $\frac{m-4}{m^2} \cdot \frac{m^2-1}{m^2-3m-4}$

$$\frac{m-4}{m^2} \cdot \frac{m^2-1}{m^2-3m-4} = \frac{m-4}{m^2} \cdot \frac{(m+\square)(m-\square)}{(m-\square)(m+\square)}$$

$$= \frac{\overset{1}{\cancel{m-4}}}{m^2} \cdot \frac{\overset{1}{\square}(m-1)}{\underset{1}{(\cancel{m-4})}\underset{1}{\square}} \text{ or } \frac{m-1}{\square}$$

Foldables™

Organize It

Summarize the steps in multiplying two rational expressions under the tab for Multiplying Rational Expressions. Then, give an example.

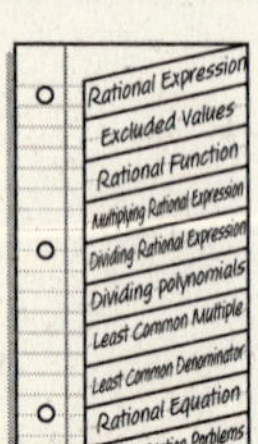

Your Turn **Find each product.**

a. $\frac{15n^2}{12m} \cdot \frac{4mn^2}{8n^5}$ ______ **b.** $\frac{2d - 10}{d^2 + 7d} \cdot \frac{3d + 21}{d - 5}$ ______ **c.** $\frac{y^2 - 64}{y^2 + 10y + 16} \cdot \frac{y^2 + 2y}{2y^3}$ ______

FOLDABLES™

ORGANIZE IT

Summarize the steps in dividing two rational expressions under the tab for Dividing Rational Expressions. Then, give an example.

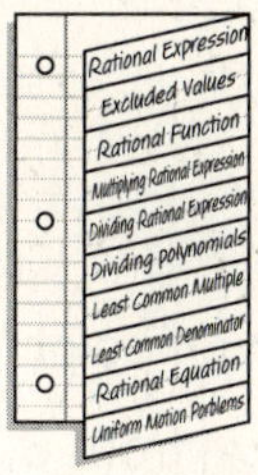

EXAMPLES

Find each quotient.

4 $\frac{15a^3}{b^3} \div \frac{5a^2}{2b}$

$\frac{15a^3}{b^3} \div \frac{5a^2}{2b}$

$= \frac{15a^3}{b^3} \cdot$ ______ The reciprocal of $\frac{5a^2}{2b}$ is ______.

$= \frac{\cancel{15}\cancel{a^3}}{\cancel{b^3}} \cdot \frac{\cancel{2b}}{\cancel{5}\cancel{a^2}}$ 5, *a* and *b* are common factors.

$= \frac{6a}{\text{______}}$

5 $\frac{6x - 6}{x - 3} \div (x - 1)$

$\frac{6x - 6}{x - 3} \div (x - 1)$

$= \frac{6x - 6}{x - 3} \cdot$ ______ The reciprocal of $(x - 1)$ is ______.

$= \frac{\text{______}}{x - 3} \cdot \frac{1}{\text{______}}$ Factor $6x - 6$.

$= \frac{\text{______}}{x - 3} \cdot \frac{1}{\text{______}}$ ______ is a common factor.

$= \frac{\text{______}}{x - 3}$

6 $\dfrac{x^2 - 36}{4x^2y} \div \dfrac{6 - x}{4x^2y^5}$

$$\frac{x^2 - 36}{4x^2y} \div \frac{6 - x}{4x^2y^5}$$

$$= \frac{x^2 - 36}{4x^2y} \cdot \boxed{}$$ The reciprocal of $\frac{6 - x}{4x^2y^5}$ is ______.

$$= \frac{\boxed{}}{4x^2y} \cdot \frac{4x^2y^5}{\boxed{}}$$ Factor $x^2 - 36$.

$$= \frac{\boxed{}\boxed{}}{(4x^2y)} \cdot \frac{4x^2y^5}{\boxed{}\boxed{}}$$ Factor ______ from $6 - x$.

$$= \frac{-\boxed{}(x + 6)}{1}$$ $x - 6$, x^2, y, and ______ are common factors.

$$= \boxed{}$$

Your Turn **Find each quotient.**

a. $\dfrac{4t^3}{s^3} \div \dfrac{12t^4}{2s}$

b. $\dfrac{p^2 - 4}{10q^2p^3} \div \dfrac{2 + p}{5q^2p}$

c. $\dfrac{2x + 10}{x - 5} \div (x + 5)$

HOMEWORK ASSIGNMENT

Page(s): ____

Exercises: ____

15–3 Dividing Polynomials

EXAMPLES

What You'll Learn

- Divide polynomials by binomials.

Find each quotient.

1 $(15x + 10) \div (3x + 2)$

$$3x + 2\overline{)15x + 10}\quad \text{quotient: } 5$$

$(-)\ 15x +$ ______

______

$15x \div 3x =$ ______

Multiply 5 and ______.

______.

Therefore, $(15x + 10) \div (3x + 2) =$ ______.

2 $(x^2 - x - 12) \div (x + 3)$

$$x + 3\overline{)x^2 - x - 12}\quad \text{quotient: } x \text{ ______}$$

$(-)\ x^2 +$ ______

$-4x$ ______

$(-)\ -4x$ ______

______

$x^2 \div x = x$

Multiply x and ______.

Subtract: $-x - 3x = -4x$;

bring down ______.

Multiply -4 and ______.

Subtract.

Therefore, $(x^2 - x - 12) \div (x + 3) =$ ______.

Your Turn **Find each quotient.**

a. $(14x - 7) \div (2x - 1)$

b. $(x^2 - 8x - 9) \div (x + 1)$

FOLDABLES

ORGANIZE IT

Give an example of a division problem using two polynomials under the tab for Dividing Polynomials. Then identify the dividend, divisor, quotient and remainder.

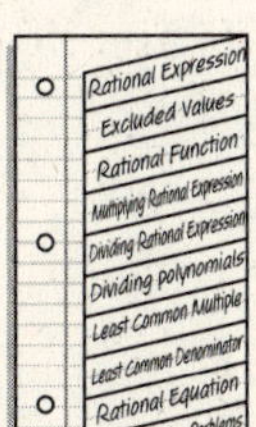

3 $(8a^2 + 14a + 9) \div (2a + 3)$

$$\begin{array}{r} 4a \quad + \quad 1 \\ 2a + 3\overline{)8a^2 + 14a + 9} \\ (-)\ 8a^2 + \square \\ \hline 2a + \square \\ (-)\ 2a + 3 \\ \hline \square \end{array}$$

$8a^2 \div 2a =$ ☐

Multiply $4a$ and ☐.

Subtract. Then bring down ☐.

Multiply 1 and ☐.

Subtract. The remainder is ☐.

The quotient is $4a + 1$ with remainder 6.

So, $(8a^2 + 14a + 9) \div (2a + 3) = 4a + 1 + \dfrac{\square}{2a + 3}$.

4 $(x^2 - 20) \div (x - 5)$

$$\begin{array}{r} x + \quad 5 \\ x - 5\overline{)x^2 + \square - 20} \\ (-)\ x^2 - \square \\ \hline 5x\ \square \\ (-)\ 5x\ \square \\ \hline \square \end{array}$$

Rename $x^2 - 20$ as $x^2 +$ ☐ $- 20$

Multiply x and ☐.

Subtract. Then bring down ☐.

Multiply 5 and ☐.

Subtract. The remainder is ☐.

Therefore, $(x^2 - 20) \div (x - 5) = \square + \dfrac{\square}{x - 5}$.

REMEMBER IT

After the divisor is multiplied by the last term in the quotient, the result is subtracted. If you prefer, you may add the opposite of the entire expression.

Your Turn **Find each quotient.**

a. $(9y^2 + 9y + 4) \div (3y + 1)$

b. $(x^2 - 32) \div (x + 6)$

EXAMPLE

5 **Find the length of a rectangle if its area is $12x^2 + 13x + 3$ square units and its width is $3x + 1$ units.**

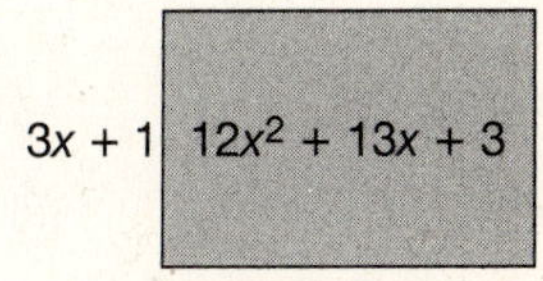

To find the length, divide the area by the length .

$$4x + 3$$
$$3x + 1 \overline{)\, 12x^2 + 13x + 3}$$

$(-)\ 12x^2 +$ Multiply $4x$ and .

$9x +$ Subtract. Then bring down .

$(-)\ 9x + 3$ Multiply 3 and .

The remainder is .

Therefore, the length of the rectangle is units.

You can check your answer by multiplying and .

Your Turn Find the length of a rectangle if its area is $5a^2 + 8a - 4$ square units and its width is $5a - 2$ units.

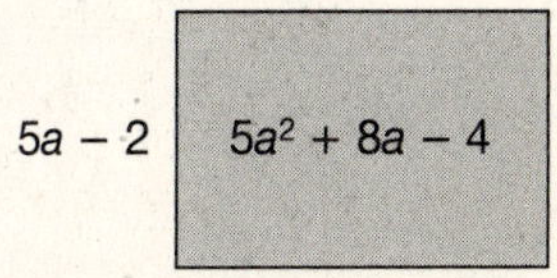

HOMEWORK ASSIGNMENT

Page(s):

Exercises:

15–4 Combining Rational Expressions with Like Denominators

WHAT YOU'LL LEARN

- Add and subtract rational expressions with like denominators.

EXAMPLES

Find each sum or difference.

1 $\frac{7}{y} - \frac{10}{y}$

$\frac{7}{y} - \frac{10}{y} = \frac{7 - \square}{y}$ The common denominator is *y*. Subtract the numerators.

$= \frac{\square}{y}$ or $-\frac{3}{y}$

2 $\frac{13a}{17} + \frac{2a}{17}$

$\frac{13a}{17} + \frac{2a}{17} = \frac{13a + \square}{17}$ The common denominator is 17. Add the numerators.

$= \frac{\square}{17}$

Your Turn **Find each sum or difference.**

a. $\frac{3}{n} - \frac{12}{n}$ $\square$

b. $\frac{5x}{11} + \frac{2x}{11}$ $\square$

EXAMPLES

Find each sum or difference. Write in simplest form.

3 $\frac{8}{15n} + \frac{4}{15n}$

$\frac{8}{15n} + \frac{4}{15n} = \frac{8 + \square}{15n}$ The common denominator is 15*n*. Add the numerators.

$= \frac{\square}{15n}$

$= \frac{\overset{\square}{\cancel{12}}}{\underset{\square}{\cancel{15}n}}$ or $\frac{\square}{5n}$ Divide by the GCF, $\square$.

REMEMBER IT

When adding or subtracting rational expressions, always check to see if your final answer can be simplified.

4 $\dfrac{5}{2p} - \dfrac{13}{2p}$

$\dfrac{5}{2p} - \dfrac{13}{2p} = \dfrac{5 - \square}{2p}$ The common denominator is $2p$. Subtract the numerators.

$= \dfrac{\square}{2p}$

$= \dfrac{\square^{4}}{\cancel{2}_{1}p}$ Divide by the GCF, □.

$= \dfrac{\square}{p}$ or $-\dfrac{\square}{p}$

5 $\dfrac{10}{3x + 1} - \dfrac{3}{3x + 1}$

$\dfrac{10}{3x + 1} - \dfrac{3}{3x + 1} = \dfrac{\square - 3}{3x + 1}$ The common denominator is $3x + 1$. Subtract the numerators.

$= \dfrac{\square}{3x + 1}$

6 $\dfrac{a - 3}{a + 3} + \dfrac{4}{a + 3}$

$\dfrac{a - 3}{a + 3} + \dfrac{4}{a + 3} = \dfrac{a - 3 + \square}{a + 3}$ The common denominator is $a + 3$. Add the numerators.

$= \dfrac{a + \square}{a + 3}$

7 $\dfrac{2x - 5}{x - 3} + \dfrac{x - 4}{x - 3}$

$\dfrac{2x - 5}{x - 3} + \dfrac{x - 4}{x - 3}$

$= \dfrac{2x - 5 + x - 4}{\square}$ The common denominator is $x - 3$. Add the numerators.

$= \dfrac{3x - \square}{\square}$

$$= \frac{\square\,(x-3)}{\square}$$ Factor the numerator.

$$= \frac{\overset{1}{\square\,\cancel{(x-3)}}}{\underset{1}{\square}}$$ Divide by the GCF, ____.

$$= \square$$

8 $\frac{12a}{2a-1} - \frac{2a-5}{2a-1}$

$$\frac{12a}{2a-1} - \frac{2a-5}{2a-1}$$

$$= \frac{\square - (2a-5)}{2a-1}$$ The common denominator is $2a - 1$. Subtract the numerators.

$$= \frac{12a - \square + \square}{2a-1}$$

$$= \frac{\square + 5}{2a-1}$$ Distributive Property

Your Turn **Find each sum or difference. Write in simplest form.**

a. $\frac{11}{14y} + \frac{1}{14y}$

b. $\frac{20}{3b} - \frac{2}{3b}$

c. $\frac{7}{x-5} + \frac{6}{x-5}$

d. $\frac{m+2}{m-10} - \frac{5}{m-10}$

e. $\frac{x+4}{x+1} + \frac{3x}{x+1}$

f. $\frac{y+1}{4y+3} + \frac{y}{4y+3}$

HOMEWORK ASSIGNMENT

Page(s):

Exercises:

15–5 Combining Rational Expressions with Unlike Denominators

WHAT YOU'LL LEARN

- Add and subtract rational expressions with unlike denominators.

FOLDABLES

ORGANIZE IT

Write two expressions and explain the steps you would take to find the LCM for the expressions under the tab for Least Common Multiple.

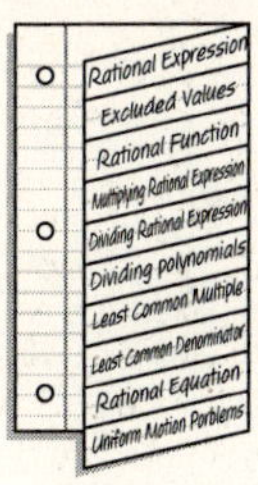

EXAMPLES

Find the LCM for each pair of expressions.

1 $12m^4n^5$, $14m^2n$

Factor each expression.

$12m^4n^5 = 2 \cdot 2 \cdot 3 \cdot m \cdot m \cdot m \cdot m \cdot n \cdot n \cdot n \cdot n \cdot n$

$14m^2n =$ ______

Use each factor the greatest number of times it appears in either factorization.

LCM = ______ $\cdot m \cdot m \cdot m \cdot m \cdot n \cdot n \cdot n \cdot n \cdot n$

= ______

2 $x^2 + 3x - 10$, $3x^2 - 7x + 2$

Factor each expression.

$x^2 + 3x - 10 = (x + 5)$ ______

$3x^2 - 7x + 2 =$ ______ $(x - 2)$

Use each factor the greatest number of times it appears in either factorization.

LCM = ______

Your Turn **Find the LCM for each pair of expressions.**

a. $15x^3y^5$, $18x^2$

b. $x^2 + 2x - 15$, $x^2 + 11x + 30$

BUILD YOUR VOCABULARY (page 318)

The least number that is a ________ of two or more numbers is the **least common multiple (LCM)**.

The least common multiple of the denominators is called the **least common denominator (LCD)**.

FOLDABLES

ORGANIZE IT

Write two rational expressions and explain the steps you would take to find the LCD for the expressions under the tab for Least Common Denominator.

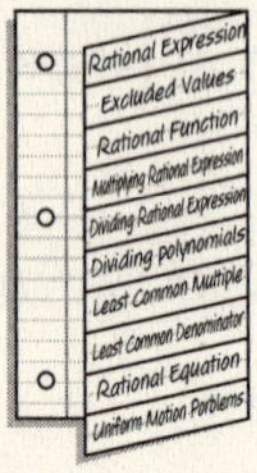

EXAMPLES

Write each pair of rational expressions with the same LCD.

3 $\frac{3}{5m}, \frac{4}{2m^2}$

First find the LCD.

$5m = 5 \cdot m \qquad 2m^2 =$ ________

LCD $= 2 \cdot 5 \cdot$ ________ or ________.

Then write each fraction with the same LCD.

$\frac{3}{5m} \cdot \frac{2m}{2m} = \frac{6m}{10m^2} \qquad \frac{4}{2m^2} \cdot$ ________ $= \frac{\text{\_\_\_\_}}{10m^2}$

4 $\frac{1}{x-3}, \frac{5x}{2x-6}$

First find the LCD.

$x - 3 = x - 3 \qquad 2x - 6 = 2$ ________

LCD = ________

Then write each fraction with same LCD.

$\frac{1}{x-3} \cdot$ ________ $= \frac{\text{\_\_\_\_}}{2(x-3)} \qquad \frac{5x}{2x-6} = \frac{\text{\_\_\_\_}}{2(x-3)}$

WRITE IT

A common denominator can always be found by multiplying the two denominators together. What are the reasons this method is not always used?

____

____

____

____

____

Your Turn **Write each pair of rational expressions with the same LCD.**

a. $\frac{2}{3a^3}, \frac{4}{5a^2}$

____

b. $\frac{n}{n+8}, \frac{3}{4n+32}$

____

EXAMPLES

Find each sum or difference. Write in simplest form.

5 $\frac{3}{4m^3} + \frac{5}{8m^2}$

Find the LCD.

$4m^3 = 2 \cdot 2 \cdot m \cdot m \cdot m$ $8m^2 = 2 \cdot 2 \cdot 2 \cdot m \cdot m$

LCD = ______ $\cdot\, m \cdot m \cdot m$ or ______

Rename each expression with the LCD as the denominator.

$\frac{3}{4m^3} \cdot$ ______ $= \frac{\_\_\_}{8m^3}$ $\frac{5}{8m^2} \cdot \frac{m}{m} = \frac{\_\_\_}{8m^3}$

Add.

$\frac{3}{4m^3} + \frac{5}{8m^3} = \frac{\_\_\_}{8m^3} + \frac{\_\_\_}{8m^3}$ or $\frac{5m + \_\_\_}{8m^3}$

6 $\frac{3}{x+6} - \frac{4x}{x^2-36}$

$x + 6 = x + 6$ $x^2 - 36 =$ ______

LCD = ______

$\frac{3}{x+6} - \frac{4x}{x^2-36} = \frac{3}{x+6} \cdot \frac{(x-6)}{(x-6)} - \frac{4x}{(x+6)(x-6)}$

$= \frac{3x - \_\_\_}{(x+6)(x-6)} - \frac{4x}{(x+6)(x-6)}$

$= \frac{3x - 18 - 4x}{\_\_\_} = \frac{\_\_\_ - 18}{(x+6)(x-6)}$ or $-\frac{\_\_\_}{(x+6)(x-6)}$

Your Turn **Find each sum or difference. Write in simplest form.**

a. $\frac{3}{10b^2} + \frac{5}{5b^3}$

______

b. $\frac{7}{x+3} - \frac{2x}{x^2 - 3x - 18}$

______

HOMEWORK ASSIGNMENT

Page(s):

Exercises:

15–6 Solving Rational Equations

WHAT YOU'LL LEARN

- Solve rational equations.

BUILD YOUR VOCABULARY (page 318)

An equation that contains at least one rational ______ is a **rational equation**.

EXAMPLES

Solve each equation.

1 $\frac{5x}{4} + \frac{2}{3} = \frac{7x}{12}$

$$\frac{5x}{4} + \frac{2}{3} = \frac{7x}{12}$$ The LCD is 12.

$$\underline{\qquad}\left(\frac{5x}{4} + \frac{2}{3}\right) = \underline{\qquad}\left(\frac{7x}{12}\right)$$ Multiply each side by 12.

$$\underline{\qquad}\left(\frac{5x}{4}\right) + \underline{\qquad}\left(\frac{2}{3}\right) = \underline{\qquad}\left(\frac{7x}{12}\right)$$ Distributive Property

$$\overset{\underline{\quad}}{\cancel{12}}\left(\frac{5x}{\underset{1}{\cancel{4}}}\right) + \overset{\underline{\quad}}{\cancel{12}}\left(\frac{2}{\underset{1}{\cancel{3}}}\right) = \overset{1}{\cancel{12}}\left(\frac{7x}{\underset{1}{\cancel{12}}}\right)$$

$$\underline{\qquad} + \underline{\qquad} = 7x$$

$$15x + 8 - \underline{\qquad} = 7x - \underline{\qquad}$$ Subtract ______ from each side.

$$8 = \underline{\qquad}$$

$$\frac{8}{\underline{\qquad}} = \frac{-8x}{\underline{\qquad}}$$

$$\underline{\qquad} = x$$

So, the solution is ______.

FOLDABLES™

ORGANIZE IT

Tell the difference between a rational expression and a rational equation under the tab for Rational Equations. Then compare the method of solving rational equations with the method of adding rational expressions with unlike denominators.

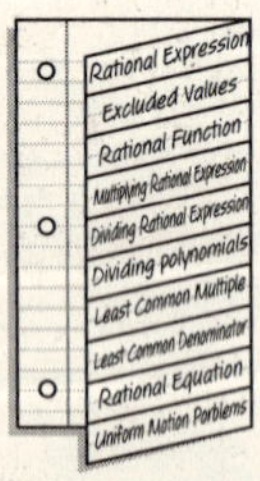

2 $\dfrac{5}{3x} - \dfrac{4}{5x} = \dfrac{26}{15}$

$$\frac{5}{3x} - \frac{4}{5x} = \frac{26}{15}$$ The LCD is 15*x*.

$$\square\left(\frac{5}{3x} - \frac{4}{5x}\right) = \square\left(\frac{26}{15}\right)$$ Multiply each side by the LCD.

$$\square\left(\frac{5}{3x}\right) - \square\left(\frac{4}{5x}\right) = \square\left(\frac{26}{15}\right)$$ Distributive Property

$$\overset{5}{\cancel{15x}}\left(\frac{5}{\underset{1}{\cancel{3x}}}\right) - \overset{3}{\cancel{15x}}\left(\frac{4}{\underset{1}{\cancel{5x}}}\right) = \overset{1}{\cancel{15}}x\left(\frac{26}{\underset{1}{\cancel{15}}}\right)$$

$$\square - \square = 26x$$ Simplify.

$$\square = 26x$$

$$\square = x$$ Divide.

3 $\dfrac{7}{x-1} + \dfrac{2x}{x-1} = 5$

$$\frac{7}{x-1} + \frac{2x}{x-1} = 5$$

The LCD is $x - 1$.

$$\square\left(\frac{7}{x-1} + \frac{2x}{x-1}\right) = \square 5$$

$$\square\left(\frac{7}{x-1}\right) + \square\left(\frac{2x}{x-1}\right) = \square 5$$

$$\overset{1}{(\cancel{x-1})}\left(\frac{7}{\underset{1}{\cancel{x-1}}}\right) + \overset{1}{(\cancel{x-1})}\left(\frac{2x}{\underset{1}{\cancel{x-1}}}\right) = \square 5$$

$$7 + \square = 5x - \square$$

$$7 + 2x - \square = 5x - 5 - \square$$

$$7 - \square = -5$$

$$7 - \square - 7 = -5 - \square$$

$$-3x = -12$$

$$x = \square$$

Your Turn **Solve each equation. Check your solution.**

a. $\frac{6n}{7} - \frac{n}{2} = 5$

b. $\frac{4}{y} + \frac{3}{2y} = \frac{11}{6}$

c. $\frac{x+1}{x-1} - \frac{2}{x-1} = \frac{x}{8}$

d. $\frac{6}{a+1} - \frac{12}{a} = \frac{-5}{a+1}$

HOMEWORK ASSIGNMENT

Page(s):

Exercises:

CHAPTER 15

BRINGING IT ALL TOGETHER

STUDY GUIDE

FOLDABLES™	VOCABULARY PUZZLEMAKER	BUILD YOUR VOCABULARY
Use your **Chapter 15 Foldable** to help you study for your chapter test.	To make a crossword puzzle, word search, or jumble puzzle of the vocabulary words in Chapter 15, go to: www.glencoe.com/sec/math/t_resources/free/index.php	You can use your completed **Vocabulary Builder** (page 318) to help you solve the puzzle.

15-1

Simplifying Rational Expressions

Simplify each rational expression.

1. $\frac{7a^2b}{28ab^3}$

2. $\frac{y(y-6)}{4y-24}$

3. $\frac{x+4}{x^2+5x+4}$

4. $\frac{9m^2-18m}{m^2+6m-16}$

15-2

Multiplying and Dividing Rational Expressions

Find each product or quotient.

5. $\frac{10x^3y}{z} \cdot \frac{xz^2}{8y^2}$

6. $\frac{3n-15}{n^2} \cdot \frac{n}{6n-30}$

7. $\frac{9ab^3}{c^2} \div 27ab$

8. $\frac{x^2-9}{2} \div \frac{x^2-10x+21}{8}$

15-3 Dividing Polynomials

Find each quotient.

9. $(12d + 30) \div (2d + 5)$

10. $(x^2 + 6x - 7) \div (x + 7)$

11. $(6y^2 + 7y + 5) \div (3y + 2)$

12. $(a^3 + 4a - 4) \div (a - 1)$

15-4 Combining Rational Expressions with Like Denominators

Find each sum or difference. Write in simplest form.

13. $\frac{7n}{10} - \frac{n}{10}$

14. $\frac{13}{9x} + \frac{5}{9x}$

15. $\frac{6y}{y - 2} + \frac{y - 1}{y - 2}$

16. $\frac{3x^2}{x - 2} - \frac{2x + 8}{x - 2}$

15-5

Combining Rational Expressions with Unlike Denominators

Find each sum or difference. Write in simplest form.

17. $\frac{2}{3x} - \frac{x+3}{9x^2}$

18. $\frac{8}{5a} + \frac{a-2}{a}$

19. $\frac{2}{n-4} + \frac{7}{n}$

20. $\frac{2x}{x-3} - \frac{9}{x^2-9}$

15-6

Solving Rational Equations

Solve each equation. Check your solution.

21. $\frac{2n}{5} - \frac{n}{2} = \frac{3}{10}$

22. $\frac{5}{6x} + 3 = \frac{7}{2x}$

23. $\frac{2}{5y} + \frac{3}{5} = \frac{y+1}{y}$

24. $\frac{a}{a+2} - \frac{8}{a-1} = 1$

CHAPTER 15

Checklist

ARE YOU READY FOR THE CHAPTER TEST?

Math Online

Visit **algconcepts.com** to access your textbook, more examples, self-check quizzes, and practice tests to help you study the concepts in Chapter 15.

Check the one that applies. Suggestions to help you study are given with each item.

☐ **I completed the review of all or most lessons without using my notes or asking for help.**

- You are probably ready for the Chapter Test.
- You may want take the Chapter 15 Practice Test on page 679 of your textbook as a final check.

☐ **I used my Foldable or Study Notebook to complete the review of all or most lessons.**

- You should complete the Chapter 15 Study Guide and Review on pages 676–678 of your textbook.
- If you are unsure of any concepts or skills, refer back to the specific lesson(s).
- You may also want to take the Chapter 15 Practice Test on page 679.

☐ **I asked for help from someone else to complete the review of all or most lessons.**

- You should review the examples and concepts in your Study Notebook and Chapter 15 Foldable.
- Then complete the Chapter 15 Study Guide and Review on pages 676–678 of your textbook.
- If you are unsure of any concepts or skills, refer back to the specific lesson(s).
- You may also want to take the Chapter 15 Practice Test on page 679.

Student Signature

Parent/Guardian Signature

Teacher Signature